Lynn M. Frock

WHY TQM FAILS AND WHAT TO DO ABOUT IT

WHY TQM FAILS AND WHAT TO DO ABOUT IT

Mark Graham Brown

Darcy E. Hitchcock

Marsha L. Willard

Burr Ridge, Illinois
New York, New York

Sponsoring editor: Jean Marie Geracie
Project editor: Karen M. Smith
Production manager: Jon Christopher
Designer: Jeanne M. Rivera
Art manager: Kim Meriwether
Art studio: Steadman-Gibson Corporate Design, Inc.
Compositor: BookMasters, Inc.
Typeface: 11/13 Times Roman
Printer: Book Press, Inc.

Library of Congress Cataloging-in-Publication Data

Brown, Mark Graham.
Why TQM fails and what to do about it / Mark Graham Brown, Darcy E. Hitchcock, Marsha L. Willard.
p. cm.
Includes bibliographical references and index.
ISBN 0-7863-0140-6
1. Total quality management. I. Hitchcock, Darcy E. II. Willard, Marsha L. III. Title:
HD62.15.B763, 1994
656.5'62—dc20 93-44771

Printed in the United States of America
3 4 5 6 7 8 9 0 BP 1 0 9 8 7 6 5 4

Preface

Much has been written recently about the failures of total quality management (TQM), as if this is just another management fad in decline, following in the grand tradition of quality of work life programs, quality circles, and the like. When they read about these problems, managers shake their heads, sighing, and then go in search of the next panacea, this time with a little less enthusiasm. Many try not to smirk too openly, as if they always knew they could out-wait this "program" as they have successfully withstood a barrage of previous management initiatives. And those in our workforce who invested their spirits and energy in TQM efforts are reminded of just how powerless they are.

We believe that this perspective is seriously flawed. After all, how can you dispute the need to delight customers and produce high-quality products and services? As an organizational philosophy, total quality management is even more critical now than it was just 10 years ago. If there has been a failure, it is not one of philosophy; it is one of implementation. And if we allow management to go in search of a replacement, we only enable them to avoid facing their own failures, thus perpetuating the search for the silver bullet.

Many TQM efforts have not yielded the expected results; we do not dispute that. However, we believe there are common causes for those failures. The purpose of this book is to explain the common mistakes that organizations make and to provide guidance to avoid those mistakes.

WHO SHOULD USE THIS BOOK

Total quality management has caught the attention of organizations in every business sector, including government, education, health care, and nonprofit. Some organizations have been working at TQM for a number of years; others are still just exploring the notion. While not a primer on TQM principles and procedures, this book is as appropriate and useful to those organizations just starting out as it is for those who are years into the process.

- It will help organizations just embarking on the TQM journey to anticipate—and thus avoid—some of the more common mistakes and pitfalls.
- It will help organizations that have stalled in their implementation to work past their obstacles and achieve new heights.
- It also will help organizations with successful quality programs to seamlessly integrate quality into their day-to-day business practices.

HOW TO USE THIS BOOK

Our intent with this book is not to prescribe a plan for implementing TQM. That is something that should be designed to the specific needs and culture of your own organization. Instead, we have focused on the most common mistakes that organizations make in their implementations. By highlighting these failings and providing guidance about how to avoid or resolve them, we hope to help readers ensure that their TQM efforts achieve their intended results.

In pursuing TQM, organizations tend to go through three identifiable phases:

Phase 1: Start up. In this first phase people at all levels of the organization struggle to learn about TQM and its principles. Early efforts generally involve implementing quality improvement projects by using the tools and techniques of TQM.

Phase 2: Alignment. In the second phase the organization discovers that quality is more than the sum of isolated improvements. To really leverage quality, organizations must align their organizational systems and practices to support quality and teamwork.

Phase 3: Integration. In the third phase the organization strives to so completely integrate TQM principles into every aspect of the organization's operations that its influence becomes invisible and automatic.

Each phase brings its own challenges and common mistakes, so we have divided the book into three parts, one for each implementation phase. The chapters within each part address the common reasons for failure at each phase. The chapters each follow the same rough outline: a description of the common mistakes and an explanation of the principles that should guide a solution. We also have tried to provide numerous examples from both the private and public sectors to inspire your own ideas. We do not attempt to supply a cookbook recipe for success, but, rather, to provide you with information to guide your efforts so you can avoid the same mistakes.

A chapter-by-chapter summary of the mistakes within each phase follows.

PHASE 1: START-UP

In the beginning, everything is new and foreign. Organizations struggle to understand and implement the basic precepts of total quality.

Chapter One Commitment: How Executives and Managers Can Support TQM

Executives sometimes fail to develop the strong commitment to TQM that is required to successfully lead the effort. In other cases, executives support the initiative but do not demonstrate their commitment with appropriate actions. Chapter 1 explains how executives and managers can effectively demonstrate their commitment.

Chapter Two Justification and Timing: How to Increase the Probability of Success

Organizations sometimes pursue TQM for the wrong reasons. Customers may demand it, competitive pressures may necessitate it, or it may be perceived as a shortcut to bottom-line improvements. Chapter 2 explores the factors that impact a TQM initiative, and it explains how to pace the implementation to increase the probability of success.

Chapter Three Education: How to Get the Most from Your Training Investment

TQM requires a significant investment in education and training. However, many organizations do not see a return on their training dollars because they implement the wrong training, or they implement the training in the wrong way. Chapter 3 explains the four stages of training and explains how training can be approached to yield positive results.

Chapter Four Results: How to Get TQM to Pay for Itself

Even though TQM should yield results in the first year, many organizations fail to see an adequate return on their TQM investment. Usually, it's because too much focus was put on short-term, cost-cutting activities that waste time, money, and other resources. Chapter 4 explains how to balance activities and results so returns from a moderate initial investment will fund future efforts.

PHASE 2: ALIGNMENT

Total quality requires more than isolated improvement efforts. To be effective the systems and structures of the entire organization must be aligned to quality.

Chapter Five Implementation Strategy: How to Avoid Making TQM a "Program" and Adding Bureaucracy

Employees in many organizations think that quality is separate from work and, in their implementation strategies, organizations do much to reinforce this perception. Typically, the effort is isolated from the organization and implemented by a hierarchy of committees. Quality is addressed by quality improvement teams (QITs), rather than the work group as a whole. Chapter 5 explains how to avoid making quality a program and how to integrate it into the organization.

Chapter Six Measurement: How to Select Meaningful Quality Measures

Organizations tend to measure what is easy to count, not what is important. Others collect too much data, too little data, or data that are misleading. Chapter 6 examines the indexes that are critical to track, and it explains how to implement appropriate measures to manage the quality effort.

Chapter Seven Appraisals: How to Redesign Your Performance Appraisal System to Support Teams

Standard performance appraisal systems often conflict with quality efforts and the need for teamwork. This chapter provides suggestions for revamping old appraisal systems and creating processes that support quality and teamwork.

Chapter Eight Rewards: How to Compensate Executives and Employees

Traditional compensation systems focus on bottom-line results. To appropriately support the TQM effort, organizations need to implement a non-traditional compensation system that strongly links quality and customer satisfaction with pay. Chapter 8 explores many of the alternative systems gaining popularity, and it explains how to design a compensation system that reinforces positive performance with pay.

PHASE 3: INTEGRATION

Integration implies that TQM has infiltrated every nook and cranny of an organization: its management style, power structure, organization design, and systems.

Chapter Nine Power Structure: How to Empower Employees

Total quality demands a radical change in the power structure of an organization. In this chapter, we explain the empowerment scale and discuss how self-directed, high-performance teams (at the extreme end of the scale) promote quality. We also describe the common barriers to implementing self-directed teams and explain how to overcome them.

Chapter Ten Management Beliefs: How to Align Philosophies and Practices

No one suffers a greater role change than managers, and with this role change comes new required competencies. In this chapter, we explain why just focusing on management style and behavior is inadequate. Management must adopt new management beliefs that will guide their actions. We explain the four basic management competencies required of TQM interventions and discuss how to help managers make the transition.

Chapter Eleven Structure: How to Design the Organization and Jobs to Promote Quality

Many quality problems are not the fault of employees but, rather, are the fault of the organizational structure. Organizations must be so structured that fiefdoms are quashed, synergies are leveraged, and innovation is encouraged. Furthermore, job design also affects quality, in that it impacts ownership, communication, and empowerment. In this chapter, we provide the principles that should guide organization and job design into the future.

Chapter Twelve Systems: How to Leverage Financial, Information, and Planning Systems

While human resource systems became significant barriers in Phase 2, additional organizational systems become hindrances in Phase 3. In this chapter, we examine how financial, information, and planning systems must be reinvented to support empowerment, quality, and customer satisfaction.

Chapter Thirteen Organizational Learning: How to Leverage Knowledge and Experience

Continuous improvement depends on an organization's ability to learn from experience and to promote experimentation and exploration. This chapter suggests strategies for creating a learning organization on three levels: organizational, leadership, and team.

Epilogue The Future: The Total Quality Community

We close the book with a glimpse into the future—the extension of TQM into the community and larger system.

Mark Graham Brown
Darcy E. Hitchcock
Marsha L. Willard

Acknowledgments

In writing a book on total quality management, we find it is impossible to thank all those who contributed to our thinking, for thankfully we work in a field where ideas are openly shared. We are grateful to all those who came before us and to our clients who allowed us into their organizations to learn with them.

We would like to personally acknowledge those individuals who granted us interviews and waded through early versions of our manuscript. These include:

Kathleen Baker, Ed Kruskamp, Deb Daniels, Doug Phillips, Bill Nix, and Katy Riding, Hewlett-Packard

Bev Bow, Emanuel Hospital

John Coné, Sequent Computer Systems, Inc.

Ron Decker, Norm Thompson

Don Forbes, Oregon Department of Transportation

Rich Gibler, Southwest Washington Medical Center

Lee Hebert and John McCabe, Monsanto Chemical Company

Kim King and Toni McConnel, Warn Industries

Karin Kolodziejski, Tektronix

Jerry Miller and Ron Parker, Bright Wood Corporation

Peter Troccoli, Precision Castparts Corporation

Pete Sinclair, US Bancorp

Rose Anne Stevenson, Boeing Commercial Airplanes

David Stout, Weyerhaeuser

Tamsen Wassell, Northwest Natural Gas

We would also like to thank Ned Hamson, editor of the *Journal for Quality and Participation,* for encouraging us to combine our many articles into a book-length work; and Lynn Berner Kilbourn, who showed enormous skill in refining portions of the final manuscript.

Finally, we thank Jean Marie Geracie of Irwin Professional Publishing for managing this project from start to finish.

M.G.B.
D.E.H.
M.L.W.

Contents

Note: The name in parentheses denotes the author of the chapter.

PART I

WHY ORGANIZATIONS FAIL DURING START-UP

Organizations embarking on their journey toward total quality will travel predictable routes and encounter predictable problems. In this first phase, organizations must acquire a basic understanding of total quality management (TQM) and its philosophy, practices, and tools. They must make strategic decisions about how and when to begin. They typically attend and conduct a number of workshops on quality topics and begin pilot projects to test the applicability of TQM principles in their organizations.

Organizations in Phase 1 are faced with a dizzying array of choices. Confronted with a growing lexicon of acronyms and techniques, executives must wade through stacks of resources, workshop fliers, and consultant brochures. They receive conflicting information about the costs and training involved. They are forced to reconcile the widely publicized TQM miracle stories with numerous reports of TQM failures in such sources as *The Wall Street Journal* and *Newsweek.* In the midst of all this confusion, they must make strategic choices about how, when, and where to get started. It's no wonder that somewhere between one-half and three-quarters of the organizations implementing TQM drop their initiatives within the first two years. Based on our experience, TQM efforts fail in Phase 1 for the following four reasons.

Lack of Management Commitment

Executives are sometimes attracted to TQM for the wrong reasons. Their customers may demand it or executives may mistake TQM as purely a cost-cutting strategy. Even if they implement TQM for appropriate reasons, they often do not know how to support the effort. Chapter One explains what executives and managers must do to demonstrate their commitment.

Poor Timing and Pacing

Sometimes organizations must experience a financial crisis before they begin TQM. While some organizations have survived this shock therapy to win the Malcolm Baldrige Award, waiting until a crisis limits the resources that can be spent on training and measurement technology. Other organizations fail because they implement TQM without identifying a compelling need. Some organizations overwhelm their employees with overlapping, competing change efforts. Chapter Two explains how to time and pace the implementation to ensure success.

Wasted Education and Training

TQM requires a significant investment in education and training. However, many organizations do not see a return on their training dollars because they implement the wrong training or implement the training in the wrong way. Eventually, executives abandon the effort for lack of tangible returns. Chapter Three explains what training approaches will yield the best results.

Lack of Short-Term, Bottom-Line Results

While TQM is a long-term organizational improvement strategy, many organizations are led to believe that they will not see any results for several years. Organizations are encouraged to focus on process, not results, as the organization's scarce resources are poured into quality activities without demonstrating results. Consequently, when the economy dips or business gets tight, TQM is dropped as a nice-to-have. Chapter Four explains how to balance process and results so returns from a moderate initial investment will fund future efforts.

These four problems are interdependent. For instance, excessive training expenses can lead to the lack of bottom-line results, which can lead to lack of executive commitment. However, with a sound plan and strategic use of training and executive time, organizations can achieve dramatic successes in Phase 1 that will help fuel excitement for Phase 2.

Chapter One

Commitment

How Executives and Managers Can Support TQM

The foundation of an effective total quality management (TQM) effort is commitment. By definition, commitment is an intellectual characteristic, a personal attribute that, like honesty, can't be mandated or imposed from outside. For you, as an executive, to successfully implement TQM in your organization, you have to believe in it—be committed to it—yourself. That is the first step. The second and equally important step is to demonstrate your belief—your commitment—to those around you.

There are no rules that, when meticulously followed, result in the establishment of a strong commitment. There are, however, guidelines, tips, and hints that can help the already committed executive demonstrate that belief more visibly and more effectively. These guidelines are discussed in this chapter, along with behaviors that indicate commitment and with several action items you can put into practice immediately.

WHY IT LOOKS LIKE EXECUTIVES DON'T SUPPORT TQM—EVEN WHEN THEY DO

When a program—TQM, EEO, SPC, any program—falls short of delivering anticipated returns, the failure often is placed at the door of the executive suite. Sometimes it's justifiable. Many executives *are* content to give quality initiatives cursory attention—so many, and so cursory, that some people believe the lack of executive commitment is the primary reason TQM fails. Others concede that, if it's not the number one reason, it's at least in the top three.

On the other hand, many executives, in both large and small companies, really believe that TQM will help make their organizations more successful. They really are committed to the total quality management effort. Most of these executives think they are doing things that adequately demonstrate their commitment. In reality, even though they are doing them for the right reasons, they often are doing the wrong things.

Typically, these managers see the implementation of TQM as something that can be and should be delegated to others. With the best of intentions, the CEO or president of the company creates a new position—often called "vice president of quality"—with the new person reporting directly to the CEO. This action is intended to symbolize the organization's commitment to quality and to demonstrate the high priority placed on the initiative. Although it does, the message is subtle. The existence of the new position communicates far more loudly that the rest of the executives, from the CEO down, have exempted themselves from involvement in the quality effort.

The intentions of executives who pursue this approach are distorted further by the criteria used to select the individual who will lead the initiative. It is rare that a highly successful executive is taken out of the line organization, or even out of another support organization, to become the focal point for quality.

Some organizations maintain that quality should be everyone's job. The disadvantage of this approach is that it is unfocused. In theory, everyone has responsibility; but, in reality, no one has accountability. Although this challenge has been successfully met in many organizations, the key is strong leadership at the top. Typically, the CEO and other executives become too busy to lead the quality journey, and leadership falters.

Many organizations opt to secure the services of a consultant. While there are many good consultants, using them to spearhead a quality initiative presents a situation similar in approach to the vice president of quality—it relieves the leaders of the responsibility to lead. The financial investment this approach entails gives many executives yet another excuse to avoid involvement. In other words, they are paying someone else to lead the effort.

WHEN ACTIONS SPEAK LOUDER THAN WORDS

Many executives acknowledge the organization's quality initiative, but they do so with gestures that are perceived as superficial. These executives might give speeches on the importance of quality initiatives, they might get together with other executives to develop a mission statement and values, they might approve a huge budget for quality related training, or they might even preside at ceremonies recognizing the commitment of other employees. But when all is said and done, these executives return to what they view as their real responsibility—the financial well-being of the company.

By their actions, and in the way they allocate their time and resources, these executives are viewed by employees as being removed from customers and suppliers, out of touch with employees, uninterested in quality data, and, therefore, uninvolved in the quality effort.

BEHAVIOR THAT DEMONSTRATES COMMITMENT

Executives who are successfully providing the leadership necessary for the implementation of total quality management, like their less-effective colleagues, give speeches on quality and, on occasion, even hand out quality buttons, coffee cups, and plaques. But these executives back up their words with actions. They literally live their commitment. There are six behavioral characteristics that usually identify executives who are effectively demonstrating their commitment to the organization's quality effort. Let's discuss these indicators of executive commitment one by one.

Time is spent with customers. One of the most powerful ways executives demonstrate their commitment to total quality management is by spending a significant amount of time with customers. Unlike the executives, particularly those in larger companies, who tend to spend most of their time on strategic and financial issues, the committed executives allocate time to focus on the issues that keep them in touch with customers and suppliers.

Suppliers are considered partners, rather than adversaries. The manner in which executives interact with suppliers is another clue to their commitment. Committed executives support a partnering relationship with suppliers, as opposed to the traditional adversarial approach.

Time is spent in the plant, on the floor, or in other areas where the day-to-day work is done. Building rapport with employees also demonstrates a commitment to quality. However, more is involved than simply being visible. Employees often are suspicious of periodic management visits that are promoted as an informal way for them to meet and interact with executives. They view the visits as an excuse for an inspection, thus negating any opportunity for positive rapport-building. The key to establishing an appropriate relationship with employees is sincerity and consistency. Those who are most successful have developed an interpersonal approach tailored to their individual management style and personality.

Time is spent attending quality related education and training courses, and in quality related committee and team meetings. One of the best ways executives can demonstrate a commitment to the quality effort is to visibly participate in training courses and in quality related committee or team meetings. The executive's participation in the full session tells employees that the course really is important.

Results in the areas of customer satisfaction and quality are reviewed, along with financial and operational results. A good measure of an executive's commitment to TQM is the amount of time spent reviewing customer satisfaction and quality data, as compared with financial and operating data. In many organizations, executives do not review quality data unless such circumstances as a product recall or a design flaw creates major problems. Executives who take a consistent interest in the quality of the organization's products or services, as well as the bottom line, demonstrate their commitment in one of the strongest ways possible.

Money and other resources are committed to total quality management. This is the truest test of commitment and, human nature being what it is, it's one of the first things employees will look for. In an economic climate where most companies are looking for ways to reduce overhead expense, many organizations scale back or discontinue their TQM effort because executives feel the cost is too high. They argue that training sessions and team meetings add to overhead expense. On the other hand, executives who are committed will view the expense as an investment in the future.

EIGHT WAYS EXECUTIVES CAN DEMONSTRATE COMMITMENT

There are several things that all executives can and should do to demonstrate support of the TQM effort. The eight action items that follow encompass the behaviors discussed above and can help executives in all types of organizations demonstrate their commitment to customer satisfaction and quality.

Action Item 1: Learn Quality Related Concepts and Skills

The executives who lead an organization on its quality journey require a great deal of training to equip them for the task. In the estimation of at least one major firm that provides TQM training, as much as 15 days of training during the first year of the TQM effort is essential. While others consider 15 days excessive, all agree the time spent in training needs to be measured in days—not hours—and there should be several.

In most organizations, where it's rare to find a vice president or a director who spends more than a day or two a year in formal training sessions, it is difficult to convince executives and upper-level managers to attend training programs. Typically, organizations devote 40 hours or more to train lower-level employees on TQM concepts and skills. Executives in these same organizations receive an "executive overview" of four to eight hours, along with a binder of support material (which most admit they never open).

For TQM to become a way of life in your organization, the ratio needs to be reversed. The executives and managers need to have more knowledge and skills than other employees. Executives need to go beyond talking about TQM. They have to apply it to their own jobs.

Incidentally, while it's a commendable start, it's not enough that executives just attend training programs. They also should *actively* participate on committees and teams that address quality issues. Many organizations form a steering committee to oversee the implementation of TQM across all locations and functions in the company. For these steering committees to be effective, they need to include top management as well as lower-level vice presidents and directors.

Tailor courses to executives' needs. The training needs of executives are different from those of the rest of the organization and, predictably, TQM training often fails when organizations adopt a "one size fits all" approach. (More information can be found in Chapter Three.) TQM training for executives should be tailored to the needs of the executive audience. Focus should be maintained on the role of executives and their responsibilities in the TQM implementation. Some topics that might be covered in TQM training for executives are:

- How to establish an infrastructure for the TQM effort.
- The role of the executive.
- Techniques for demonstrating commitment.
- How to link TQM with other performance measures.
- How to find time for TQM-related activities.
- How to evaluate a leadership approach.
- What executives can do to encourage middle-management support.

Action Item 2: Obtain One-on-One Coaching

Attendance at TQM training/education courses is a necessary first step that managers at all levels need to take. The next step is to obtain one-on-one coaching on the application of TQM. This approach to managing is relatively new, and it is very different from what most managers and executives have learned and practiced over the years. Management used to mean directing, monitoring, and controlling employee performance. The new approach involves inspiring employees, leading them, and empowering them to make their own decisions and solve their own problems.

It takes more than classroom training to facilitate this change in management style. It requires individualized coaching and feedback to ensure that required behavior changes occur. Even for executives who believe in TQM and who want to show their commitment, reading a book, such as this one, or attending a course does little to change behavior. The changes are made more quickly when executives have access to experienced executives who can coach them, by pointing out behavior that is inconsistent with the TQM approach and by suggesting alternative behaviors.

Who coaches the chief? TQM coaches can be either outside consultants or internal professionals with a detailed knowledge of TQM and of the coaching/consulting skills required for the job. A major manufacturer of military jet aircraft uses internal professionals. The instructors of quality related courses spend about a third of their time teaching in the classroom and two-thirds coaching lower-level managers and hourly employees. This situation, while commendable in effort, points out a common oversight—no one is coaching the executives.

The solution could be to use outside consultants. For the most part, these outsiders have more credibility with executives than the inside person who might be three or four levels below them. The disadvantage, aside from the cost, is that the consultant is not on-site every day to work with the executives. For many organizations, the best approach is a combination of outside and inside people to serve as coaches.

Action Item 3: Review Quality and Customer Satisfaction Data Regularly

Giving quality and customer satisfaction data the same high priority placed on financial data is a powerful—and a far-reaching—way to demonstrate a commitment to quality. When the CEO takes an interest in the data, it also raises the priority of quality among all executives and managers in the organization.

In most companies, top executives attend monthly or even weekly meetings. The two major topics usually discussed are a review of performance data and major issues or problems relating to the company's products, services, or operation. The data reviewed in these meetings are usually financial data, including sales and operating expenses, and operating data such as production numbers, inventory turns, and productivity.

A third item should be added to the agenda: quality and customer satisfaction data. The quality data should include internal data collected on indexes, such as defects or service failures. The customer satisfaction data should include both hard and soft measures collected from customers. (Soft measures are those from customer surveys or customer interviews, designed to elicit input about the quality of products or services. Hard measures are indexes of customer's buying behavior. For example, the number of repeat orders received from first-time customers is a hard measure of customer satisfaction.)

Internal and external customer satisfaction. As simple as it may sound, introducing a review of quality and customer satisfaction data into executive meetings presents at least one problem. Most of the executives are not directly accountable for the results the data represent. The vice presidents of finance, law, R&D, engineering, procurement, facilities, human resources, marketing, planning, and other support functions are not responsible if the quality of the company's products is poor, or if customers are dissatisfied with service. Therefore, it also is important that the CEO ask each of the organization's support functions for data on levels of internal customer satisfaction. More information on measuring internal and external customer satisfaction appears in Chapter Six.

Others note what executives notice. Others in the organization pay attention to what executives review. When the CEO or president, or both, wants to see quality and customer satisfaction data, executives throughout the organization almost immediately pay more attention to these measures. It's relatively easy to start an initiative like this—but difficult to keep it going. It takes a strong belief that the continual delighting of customers with quality products and services will lead to financial results. Although enough companies have taken this path to establish that the risks are minor, it still requires a leap of faith for many executives to forego short-term financial results to achieve better levels of customer satisfaction in the long run.

Action Item 4: Establish Reasonable Quality Goals

In recent years, it has become fashionable to set what are called "stretch goals." The purpose of arbitrary stretch goals is to encourage employees to look at new ways of doing things and to change processes to meet the goals. Motorola, one of the first companies to win a Baldrige Award, established a stretch goal it calls "Six Sigma." To achieve this goal, the company must have no more than 3.4 defects per million products.

The goal was determined as the result and an analysis of Japanese competitors that achieved fewer than 3.4 defects per million products. This is the correct way to set stretch goals—based on an analysis of what is possible.

Stretch goals can be beneficial and often provide the motivation for remarkable gains. However, they cannot be set by arbitrarily picking numbers. It is important that stretch goals, like those set by Motorola, are based on research that says someone, somewhere in the world, has achieved the performance levels specified in the goals. If employees perceive the goal as unrealistic and unreachable, they become frustrated and give up trying to achieve it.

Executives also should avoid merely copying what other organizations are doing because they like the approach. Executives for an electronics manufacturer were proud of their company's "Six Sigma Program." When asked what six sigma meant, they explained that it meant 3.4 defects per million products. The follow-up question, however, revealed that the company currently had somewhere in the neighborhood of 800 to 900 defects per million products. This company admired Motorola and had decided to adopt the six sigma goal because it sounded good. Not much thought was given to whether the goal was appropriate for the company's business or its products.

Action Item 5: Talk about TQM Efforts with Employees

Management by objectives and results is becoming passé. Management by walking around and spending time performing front-line jobs is taking its place. Successful companies have found that managers are more effective if they spend a good percentage of their time walking around where the real work of the company is done. Spending time with employees and asking about quality improvement efforts is yet another way to demonstrate commitment.

Executives who excel in this arena spend several hours a week visiting different areas of the facility they manage. They chat with people about quality improvement efforts in their jobs and areas. They look at graphs and ask questions about anomalies they see in the data. They discuss how the employees arrived at the root cause of a problem they were trying to solve. Frequently, they ask an important question about the results: How did you do that?

Executives are careful not to intimidate employees or make them feel stupid because they can't answer their questions. The atmosphere they create provides employees with an opportunity to explain their quality improvement projects, and it gives the executives opportunities to praise employees for their accomplishments.

This approach can be very successful for some executives. Others who try to manage by walking around meet with unsatisfactory results. The guidelines that follow can help ensure success.

Be sincere. When dealing with people, executives should choose an approach that is comfortable and natural, one that does not force them to display a sincerity they do not feel. If interest and enthusiasm are not genuine, or appear forced, employees will not only see through it, they'll distrust the executive and automatically be suspicious of his or her presence. It also is wise to keep in mind that trust is earned and, like interest in a savings account, it doesn't accrue overnight. It takes time to convince employees that executives are there because they are interested, not because someone did something wrong.

Vary the time and frequency of visits. After reading several management books, one manager visited the plant every day from 2:00 P.M. to 2:30 P.M. to praise his employees. This manager thought he was following the advice in the books and wondered why the approach wasn't getting the positive results he expected. The reason was that employees figured out the game, and when they did, the manager's visits lost much of their impact. This circumstance can be avoided if the executive appears to be in the area for another reason and just happens to notice a graph or just happens to talk to a person. With a casual approach, executives don't look like they are "making the rounds," talking to everyone about their quality improvement efforts.

Look for opportunities to give positive feedback. Employees in many organizations have learned through experience that an executive in the plant or work area means someone's in trouble. This attitude is countered when executives make all interactions as positive as possible. For example, when reviewing performance, executives should praise improvements, even if these are slight and even if results fall short of goals. Even mild negative feedback has far-reaching results: Everyone dreads the executives' visits, afraid that they may be the next one to be put on the spot.

Action Item 6: Implement Your Own Quality Improvement Projects

Executives who identify opportunities and problems within their own areas and who work to improve key performance measures set a powerful example. Many times, the results surprise even them. Motorola's chief financial officer, Ken Johnson, is an especially good example. Initially, he was reluctant to get involved with TQM because he didn't see how it applied to a finance function. At the urging of other executives, however, Johnson decided to try it. He chose a troublesome problem that always had created havoc in the finance department—the closing of the books—a monthly event that required a great deal of overtime and caused the staff a great deal of stress. When Johnson and his staff used the quality improvement tools and techniques they'd learned in Motorola's training courses, they simplified the month-end closing process significantly. Now, overtime is not required, and the process is completed with about half the labor hours and certainly half the grief that it used to entail. The positive results of the effort, led by the chief financial executive, prompted others in the department to initiate their own process improvement projects.

Action Item 7: Allocate Appropriate Resources for Total Quality

Many executives are strongly committed to the quality effort—until they see the price tag. Admittedly, in terms of time as well as of money, the price tag for total quality management is high, but the payoff is much greater than the investment. Nearly every company that has invested in TQM will testify that it's a wise investment that pays much more than it costs. In fact, Globe Metallurgical, a 1988 Baldrige winner, claims its TQM effort returned 40 dollars for every dollar it invested.

How to recover the investment. Total quality management can be implemented without a huge up-front investment that could take three to five years to recover. It is possible to get a good return on investment during the first year and every year thereafter. For example, the savings that result from team projects initiated during Phase 1, and the additional sales earned through consistently delighting customers, generate the revenue to pay for the training, team meetings, and recognition programs that must be budgeted initially.

As the effort gains momentum, the cost also is offset by a reduction in rework, scrap, and other expenses associated with nonconformance to requirements.

Determine the benefits. It's unrealistic to expect executives to approve a budget for a major investment without a clear way to track the return that the investment brings. Despite this, many organizations neglect to project benefits and determine ways to track them. Most know how much it costs, but few know whether the company has benefited from the effort. Often it's because results can be difficult to quantify. To illustrate, if customer satisfaction scores are increased from 3.2/5.0 to an average of 4.5/5.0, what is the increase worth to the company? To find out, some companies statistically predict the future buying behavior of customers, which enables them to calculate the benefit of these types of results.

How much is enough? Money, time, and people are the three most important resources the TQM effort requires. When allocating these resources, remember that TQM is not a program or annual initiative. Rather, it's a massive cultural and operational change that requires several years to implement.

A common mistake executives make is to assume that TQM can be implemented just by doing things differently. Or they want to know how it can be done in less time, for less money, and without adding headcount. The answer is, it can't. TQM requires substantial resources to implement, and there is no magic formula to help determine how much of each of the three resources is required.

You can gain insight, however, by talking to other executives at companies that have successfully implemented TQM. Find out what kind of resources they have spent, and if they originally had allocated enough (most don't). You'll also find that, in most cases, the costs are high—but so are the benefits.

Allocate your time. For most executives, it's much easier to allocate funds or assign staff members than it is to personally spend time on quality related activities. The fact is, though, that the time executives devote to quality related activities is as important or more important than the money and staff they allocate.

When it comes to allocating their own time, executives are unsure how much time should be spent, but it's almost always more than they would expect. Many executives believe they're doing a good job demonstrating their commitment if they spend four to eight hours a month on quality related activities, such as attending team meetings, developing plans, and recognizing employees for quality related achievement.

While this is a good start, it is far from what many other companies are doing. According to Bonnie Soodik, vice president and general manager of the Quality Systems Division of McDonnell Space Systems:

> Senior executives spend about 70 percent of their time on quality related activities. This includes six hours a week in executive council meetings, a full day of planning off-site each month, giving and receiving training, conducting roundtables, having open-door meetings with employees, and doing "walk around management" by visiting the floors. [1]

Action Item 8: Use the Best Measurement Technology Available

Total quality involves more than just measuring and improving quality. It means measuring and improving all aspects of organizational performance, including:

- Financial measures.
- Operational measures.
- Employee satisfaction measures.
- Customer satisfaction measures.

Data are only as accurate as the methods and tools used to collect them. Poorly maintained or out-of-date measurement devices and instruments yield inaccurate data that can cause executives to make faulty decisions. One of the major themes of TQM is the use of good data to track performance and aid in making decisions. In fact, accurate measurement of all important aspects of performance is an absolute requirement for improvement.

In the past 10 years, major strides have been made in the technology available to measure aspects of product quality such as color, dimensions, tolerances, and structural characteristics. One of the executive's responsibilities is to make sure the organization has the latest and best measurement technology that it can afford.

Priorities for spending on measurement equipment should be based on an analysis of customer requirements and needs. Devices to measure the factors that are most important to customers should be highest on the priority list. Keep in mind that the production area of a factory, or the service delivery area in a service company, are not the only places where performance should be measured. About two-thirds of quality problems in products and services are caused in the engineering or design phase, so it makes sense to collect key data in these areas as well. More information on how to design a good measurement system, and how to measure functions such as R&D, marketing, and engineering, appears in Chapter Six.

THE KEY TO EXECUTIVE COMMITMENT

The key to executive commitment is to spend time on the right activities. Before attempting to change how executives spend time, it might be worthwhile to track where they currently spend their time. The Quality Council, an informal group of CEOs from large American corporations, has developed a set of seven categories for executives to use in tracking their time. These seven categories and the council's suggestions on the percentage of time CEOs should devote to each are outlined in Figure 1–1.

One of the members of the Quality Council is Corning Glass CEO James R. Houghton. Corning has been heavily into TQM since 1986 and has been a finalist for the Malcolm Baldrige Award. In 1989, Houghton used the seven categories defined by the Quality Council to track his time. He found he worked a total of 2,462 hours, 1,483 of which were in what he calls "value-added activities." The nonvalue-added activities included things like travel time and unscheduled office hours.

FIGURE 1–1
Key CEO Responsibilities

	Responsibilities	*Ideal Percent of Time*
1.	Provide strategic direction, planning, and customer focus.	10–20%
2.	Monitor and evaluate operations.	5–20
3.	Organize, utilize, and develop the management team.	10–15
4.	Create appropriate environment and value systems that stimulate the morale and productivity of the workforce and leadership.	5–15
5.	Provide external representation and stay abreast of influential events and trends.	5–20
6.	Maintain positive relationships with the board of directors and shareholders.	5–10
7.	Attend to personal development and care.	5

Source: David A. Garvin, "How the Baldrige Award Really Works," *Harvard Business Review,* November–December 1991, p. 90.

He spent a total of 492 of his value-added hours—or about one-third of his time—on quality related activities. While this is much less than the 70 percent that McDonnell executives claim to spend, it is still a substantial portion of time. Houghton explains that it is appropriate for him to spend less time on quality related activities because the quality effort at Corning is in the maintenance and continual improvement mode. When a company is just getting started with TQM, executives need to spend a greater percentage of time on quality related activities. [2]

CONCLUSION

By virtue of their leadership positions, executives play a key role in the TQM effort, and a clear demonstration of commitment is especially critical. Too often, however, they do not adequately demonstrate their commitment, substituting ineffective approaches for visible actions. In this chapter, we identified some of these inappropriate behaviors and suggested a number of actions that executives can take to demonstrate their commitment.

To successfully implement these activities, executives will need to develop a strategy that is compatible with their personal style. For most, it means a major change in the way they manage and make decisions, but the results will be worth the effort. Executives will find themselves proactively leading the quality effort, rather than being swept along in its wake.

Tips, Tools, and Techniques TANGIBLE RESULTS OF EXECUTIVE COMMITMENT

Moore Business Forms

One company that goes further than simply training its executives about TQM is Moore Business Forms. Moore executives not only attend the TQM training, they conduct the TQM courses for their employees. Each executive teaches at least one session of several different TQM courses. When the executives train their direct reports, those managers, in turn, train their direct reports. The chain continues until employees at all levels have been trained. This approach, which permits tailoring courses to the needs at each level, ensures that everyone receives the knowledge and skills. Often the best way of getting someone to learn something is to tell the person he or she has to teach the subject.

This approach to training, which also is used by Baldrige Award winner Xerox Corporation, is very expensive, but there is no better way to get everyone involved in TQM training and put the responsibility for its implementation on the shoulders of management at all levels.

TRW

Employees of TRW Corporation were skeptical when the executives implemented TQM. Many dismissed it as "another program of the month." When TRW Corporation decided to use the Malcolm Baldrige Award criteria as a standard to evaluate its progress in the initiative, and all executives attended a day-long workshop to help them understand and interpret the award criteria, employees began to take the effort seriously. Never before had the president and all the executives from around the company come together for a workshop. The executives' attendance at this workshop sent a powerful and positive message to employees about the seriousness of TQM.

Coldwell Banker

Coldwell Banker Relocation Services, a small division of the real estate giant, is beating out most of its competitors in the industry by focusing attention on customer satisfaction data. The company, which provides relocation services to large companies, such as GTE and IBM, surveys relocated employees to determine their levels of satisfaction. The president and executive vice president of Coldwell Banker Relocation Services spend four to six hours a week reviewing the surveys and adding their own comments and input to the survey forms. This is, incidentally, much more time than they spend reviewing financial data. As a result of their interest, the importance of customer satisfaction filters through the organization, especially when employees get back surveys with handwritten notes from the president of the company.

SUGGESTED READING

Boyett, Joseph H., and Henry P. Conn. "What's Wrong with Total Quality Management?" *Tapping the Network Journal,* Spring 1992, pp. 10–15.

Brown, Mark Graham. "Commitment . . . It's Not the Whether, It's the How to." *Journal for Quality and Participation,* December 1989, pp. 38–42.

"Deploying TQM and Empowerment at McDonnell Douglas Space Systems Co." *Commitment Plus,* October 1992, pp. 1–4.

Garvin, David A. "How the Baldrige Award Really Works." *Harvard Business Review,* November–December 1991, pp. 80–93.

Ginnodo, William L. "How Do You Get Executives to Really Lead Quality Improvement?" *Tapping the Network Journal,* Winter 1991, pp. 7–10.

Chapter Two

Justification and Timing

How to Increase the Probability of Success

Total quality management almost always thrives in start-up companies. When these organizations are formed, systems to focus on customer satisfaction and quality are integrated into the organizational structure to such a degree that TQM *is* the organizational structure. Companies facing near-death experiences also have been able to implement TQM with a high degree of success. Threatened with financial failure, these companies are motivated to make massive system and cultural changes that result in major improvements in customer satisfaction and financial performance.

Not surprisingly, organizations in these two situations, start-up and near-death, seem to have the highest rate of success with their TQM implementation. Companies that typically struggle are those that perceive themselves to be successful, established, stable, and financially secure. In these types of companies, it is often difficult to obtain the commitment and motivation to drive the cultural changes that TQM requires. In this chapter, we will explore the factors that impact an organization's likelihood of success with TQM, and we'll present suggestions on how to schedule the timing of the TQM effort to enhance the probability of success.

HOW "WHY" AND "WHEN" CAN KILL THE IMPLEMENTATION

In the last decade literally thousands of organizations in the United States have initiated a total quality management effort. For each, there were underlying circumstances that influenced the decision to implement a TQM culture—as well as factors that dictated the time at which it would begin. Whether the cultural transition proceeded smoothly, or whether it floundered or failed, often was directly related to the circumstances that led to the implementation and the degree to which they impacted its timing. The following discussion addresses several of the factors that hamper a successful TQM implementation.

THE WRONG REASONS

For any organization embarking on a quality journey, the obvious question is: Why is the organization trying to implement TQM? The answers usually are variations of several common themes, all of which are the wrong reasons to initiate a TQM initiative. For example, many organizations pursue TQM because their major customers encourage them to do so. Some want to win the Baldrige Award, thinking that will give them an edge over their competition, while others want to prevent their competition from getting an edge over them. In still other organizations, TQM is implemented at the instigation of a CEO who read a book or heard a speech and thinks it's a good idea.

To please customers, to enhance competitive position, or to follow along with the crowd are all poor reasons to get involved with TQM. The best and only reason to engage in a TQM implementation is to improve the organization's performance in all areas: financial results, customer satisfaction, and employee satisfaction.

The Shortcut to Quality

Companies like those discussed in the above paragraphs are willing to attempt a TQM implementation to secure what they consider to be the benefits. Other organizations hope to acquire the benefits with nothing more than the illusion of TQM. For example, one company sought the services of a consultant to help develop a brochure that described the company's total quality management program. The brochure was to be a response to several major customers, and numerous potential customers, who had asked for information about the company's quality program. The flaw, however, was that the company did not have a TQM program—and had no interest in starting one. The company simply wanted the brochure—a marketing piece—and presumed a "quality consultant" would know what information it should contain.

While situations like these are certainly not common, a number of companies, to one degree or another, are looking for a shortcut to TQM, or a quick fix that will provide most of the benefits with less of the expense or effort. In organizations that adopt this approach, TQM is destined to fail. It is a long-term and cost- and effort-intensive initiative—and there are no shortcuts or quick fixes.

Too Little Too Late

Most of the concepts and tools of total quality management are very simple and easy to understand. Consequently, many executives wonder why it costs so much money to implement. Many believe the only requirements are a modest training program, a quality steering committee, several teams, and a few surveys and benchmarking studies. As a result, the "too little too late" phenomenon often occurs in organizations where the full impact of a TQM implementation is not understood—or is misunderstood. The CEO of a 75-year-old corporation is a prime example.

Although this company was always considered to be successful, it was always rated as number two or three in market share, compared to its largest competitors. The compelling force for a TQM effort was the slow erosion of market share, combined with demands from major customers that TQM systems be in place. The CEO was willing to implement TQM, but he wanted to do it in four years or less, without spending much money or time. He imagined it would only involve making a few minor changes and developing more surveys to measure customer satisfaction.

Many organizations are serious about their TQM initiative and devote significant amounts of their resources to the implementation—but this, too, is often too little too late. For example, companies like General Motors, IBM, and McDonnell Douglas started their quality journey many years ago; but the situation these companies are in today is probably an indication that the effort was not intense enough and not soon enough. It is likely that these companies, and others like them, could not change approaches fast enough to avoid financial trouble.

Incidentally, even though huge corporations like those just mentioned often lack a unified TQM approach, it is common to encounter pockets of excellence in autonomous plants or business units. For example, General

Motors' Saturn Division and IBM's AS/400 plant in Rochester, Minnesota, have both demonstrated their ability to achieve high levels of customer satisfaction and financial success with approaches based on TQM principles.

FACTORS THAT DETERMINE SUCCESS

The odds that an organization will successfully implement total quality management are two to one—against it. Based on our experience, two-thirds of all TQM implementations are dropped or fail altogether. However, organizations armed with knowledge of the factors that have the biggest impact on their success can take steps to influence or change those factors. The following discussion examines the five factors—threat, commitment, plans, progress, and strategy—that play a critical role in the successful implementation of total quality management.

Threat. Of the five factors that impact the probability of a successful TQM implementation, threat is the most powerful. If business is good and customers are happy, it is much harder to implement the changes that a TQM effort demands. Although organizations have successfully implemented TQM with no perception of immediate threat, the presence of a threat seems to make it far more likely that TQM will succeed. Typically, it is relatively easy to generate interest in the changes the implementation demands in organizations that have had a near-death experience. For example, Xerox, the inventor of photocopy technology, found Canon and other Japanese competitors could sell copier machines for close to what it cost Xerox just to manufacture its own. Under the leadership of CEO David Kearns, the company implemented a TQM approach that is credited with turning the situation around and restoring the company's financial health.

Often, an organization's current financial results do not signify a threat. However, trends may indicate otherwise. To illustrate, many defense contractors today are showing declines in sales, and they predict that long-term demand for products will diminish. Yet, employees in many of these firms sense no threat or urgency to change. Workloads include overtime, and there is no feeling that jobs are in jeopardy. Employees and managers in many firms facing longer-term threats often feel invincible, even though in reality they are not. For these companies, the time to begin a quality journey is before customers and market share begin to erode.

Commitment. The second most important factor affecting the success of an organization's TQM initiative is the degree of commitment that employees have to the effort. Although commitment is strongly linked to threat, the presence of a threat is not required for commitment.

Commitment to a change strategy like TQM is not a yes or no issue. It is best understood in the context of a scale that ranges from 0 to 100 percent (see Figure 2–1). The employee at the 100 percent level is committed to TQM both intellectually and emotionally—in words, thoughts, and deeds. Those at the 75 percent level have a strong intellectual commitment to TQM. They know it works and will help the company become more successful. However, the person at this level is not committed emotionally and still has difficulty aligning day-to-day behavior with the values of TQM. In a stressful situation, the 75 percent committed person reverts to old behavior patterns that may be contradictory to TQM.

At the 50 percent commitment level, a person shows real compliance with TQM, but no real commitment on either an intellectual or emotional level. This person goes along with the effort, because it's expected of him or her. People at the 25 percent level support the effort in public situations, but in private they are cynical. They feel TQM is another short-lived management program that will pass in a year or two. The employee at the 0 percent level is openly noncompliant. He or she believes TQM is a waste of time and verbalizes the belief. People with 0 percent commitment often are dangerous to the effort, because they are sometimes among the most successful and best performers in the company.

FIGURE 2–1
Scale of Commitment

100%	*75%*	*50%*	*25%*	*0%*
Committed on both intellectual and emotional levels.	Strong intellectual commitment; little emotional commitment.	Compliance only.	Lip service.	No compliance or commitment.

Plans. Of the five factors that affect the probability of success, this has the least impact. The assessment of current strengths and weaknesses and the development of a plan to address them is less important than other factors because a plan, by itself, does little to guarantee a long-range commitment.

Before a long-term plan for the implementation of TQM is developed, needs should be assessed against established criteria. One approach is to use the criteria for the Malcolm Baldrige National Quality Award as the basis for the assessment. Such companies as Northrop, TRW, Roadway Express, and others complete mock applications and score them as Baldrige examiners would. The scores, along with the strengths and areas for improvement that are identified, serve as a baseline for setting goals and developing plans for improvement. Other companies use ISO 9000 criteria to audit or assess quality improvement efforts. Whatever the criteria used, they should be comprehensive and well accepted.

Organizations that have a three- to five-year plan are more likely to stay with the TQM effort for the duration. Companies that have a detailed plan for TQM implementation tend to understand how the individual initiatives—for example, re-engineering, ISO 9000, benchmarking—fit together to form a TQM strategy. The plan identifies key goals and strategies to achieve the goals.

The plan, however, should not be too aggressive. Many companies make the mistake of trying to do too much too quickly, which is as counterproductive as doing too little over too long a time. Also, the plan should set goals that focus on results, rather than on the activities that will yield the result. For example, rather than set goals that specify numbers of employees that should be on teams, set goals that relate to the activities the teams are performing.

Progress. The fourth factor that impacts the probability of success with TQM is the amount of progress made in its implementation. The further along an organization is, the less likely the effort will be dropped.

The TQM implementation has three phases, as outlined in the beginning of this book. These phases are:

Phase 1: Start up.

Phase 2: Alignment.

Phase 3: Integration.

Most American companies are in Phase 1 of the implementation, which involves building awareness of TQM and training employees in basic skills, such as problem solving and teamwork. Teams also are established in Phase 1, and team projects begin to produce results.

In Phase 2, macro changes are implemented to further the integration of TQM as a way of life in the company. Major systems, such as information/databases, compensation, and recruiting and selection, are critiqued and redesigned. Until the fundamental systems used to deal with customers, suppliers, and employees are changed, TQM remains a short-lived program with short-lived results.

The principles of TQM are integrated into the organization on a day-to-day basis in Phase 3. When organizations are in Phase 3, TQM is completely invisible.

Strategy. The fifth factor that appears to influence the success of a TQM effort is the strategy employed in the implementation. For example, companies that create an artificial organization, consisting of a hierarchy of committees to oversee the implementation, employ a strategy that often results in failure.

Another critical dimension of strategy is its deployment across the organization. To be effective, strategy must involve all functions, categories, and levels of an organization.

CALCULATING YOUR CHANCES FOR SUCCESS

Based on the five factors just discussed, an organization can calculate the probability that its TQM implementation will be successful. The survey shown in Figure 2–2 is designed to help identify what factors need to be strengthened to provide the greatest likelihood of success.

The survey consists of a series of 40 statements about an organization's practices. The statements are divided into five categories to correspond to the five factors discussed above. Each of the categories is weighted according to each factor's importance in determining the success of the TQM effort. The number of questions in each of the categories corresponds to the percentage of weight given to the factor.

Threat:	30 percent.
Commitment:	25 percent.
Plans:	10 percent.
Progress:	15 percent.
Strategy:	20 percent.

Respond to the statements in each category by indicating the extent to which they are true for your company. A 5-point scale is used, ranging from 5 = strongly agree to 1 = strongly disagree.

FIGURE 2–2
Identifying Needed Strengths: A Survey

	5	*4*	*3*	*2*	*1*
	Strongly Agree				*Strongly Disagree*
Threat					
1. Our market share has been shrinking over the last few years, while the market share of some key competitors has grown.	5	4	3	2	1
2. If we don't make some dramatic changes to improve the quality of our products/services, we may be out of business in five years.	5	4	3	2	1
3. We have lost several major customers or accounts over the last few years because of quality related issues.	5	4	3	2	1
4. Customers perceive our products/services to be inferior to some of our competitors' products/services.	5	4	3	2	1
5. Most employees here are concerned about keeping their jobs because the company's products/services are not as good as they used to be.	5	4	3	2	1
6. The executives in this organization are very frightened and are willing to take some major risks to get this company back on track.	5	4	3	2	1
7. We have studied some of our competitors' products/services, and they are of better quality than ours.	5	4	3	2	1
8. Our sales and profits have declined over the last few years.	5	4	3	2	1
9. Our problems with quality are not limited to a few products/services, but they include most of our products/services.	5	4	3	2	1
10. Our quality performance record shows that we are getting worse, not better.	5	4	3	2	1
11. We no longer feel our company is invincible.	5	4	3	2	1
12. We can't afford to spend three to five years implementing an improvement strategy—we need to do something now!	5	4	3	2	1
Subtotal					______

FIGURE 2–2
(continued)

	5	4	3	2	1
	Strongly Agree			*Strongly Disagree*	
Commitment					
13. Our CEO is completely committed to TQM on both an intellectual and emotional level.	5	4	3	2	1
14. Our CEO demonstrates commitment to TQM on a day-to-day basis through behavior and decisions.	5	4	3	2	1
15. Every one of the executives reporting to the CEO is as committed to TQM as the CEO.	5	4	3	2	1
16. The executives demonstrate commitment to TQM by the amount of time they spend with customers, suppliers, and employees.	5	4	3	2	1
17. The executives demonstrate commitment to TQM through the amount of resources they devote to this effort (people, budget, and the like).	5	4	3	2	1
18. The number one priority in the company is improving customer satisfaction—it's even more important than sales or profits.	5	4	3	2	1
19. All of our middle managers are committed to TQM on both an intellectual and emotional level.	5	4	3	2	1
20. Employees in all functions and at all levels are strongly committed to TQM in their heads and hearts.	5	4	3	2	1
21. Managers and executives who refuse to support TQM will not be allowed to stay in our organization.	5	4	3	2	1
22. Layoffs or other cost-saving measures have not, and will not, cause us to cut back on our efforts to improve quality and customer satisfaction.	5	4	3	2	1
Subtotal				______	
Plans					
23. Our organization has conducted a thorough overall assessment using well-established criteria, such as ISO 9000 or Baldrige.	5	4	3	2	1
24. Our organization has developed a detailed long-term plan for the implementation of TQM.	5	4	3	2	1

FIGURE 2–2
(continued)

	5	*4*	*3*	*2*	*1*
	Strongly Agree				*Strongly Disagree*
25. The plan for implementation of TQM is comprehensive and thorough, but it does not attempt to do too much too quickly.	5	4	3	2	1
26. We have developed goals for key results, such as customer satisfaction and quality, rather than goals for activities, such as number of teams or number of training programs conducted.	5	4	3	2	1
Subtotal					______
Progress					
27. TQM is not a program or an initiative in our organization. It is the way we run our organization on a day-to-day basis.	5	4	3	2	1
28. We must make major changes to the systems in our organization to help facilitate the implementation of TQM.	5	4	3	2	1
29. We have been working hard to implement TQM for at least three years.	5	4	3	2	1
30. The problem-solving teams we formed have achieved results that are at least double the amount of the organization's investment.	5	4	3	2	1
31. Every employee in our organization has been trained on the basic skills needed to improve quality and customer satisfaction in their own jobs.	5	4	3	2	1
32. We have involved our customers and suppliers in our quality improvement efforts.	5	4	3	2	1
Subtotal					______
Strategy					
33. We have not created a hierarchy of committees that are responsible for implementing TQM.	5	4	3	2	1
34. The implementation of TQM is the responsibility of all managers and employees.	5	4	3	2	1
35. The executives in this organization are leading the TQM effort and have not delegated this responsibility to a quality VP or a consulting firm.	5	4	3	2	1

FIGURE 2–2
(continued)

	5	4	3	2	1
	Strongly Agree			*Strongly Disagree*	
36. TQM is being implemented in all the organization's business units and facilities.	5	4	3	2	1
37. There is a clear relationship between our long-term business goals and our TQM strategies.	5	4	3	2	1
38. TQM is being implemented in all support functions, such as finance, human resources, and the legal department.	5	4	3	2	1
39. We have made changes in our organization structure to facilitate empowerment and implementation of TQM.	5	4	3	2	1
40. Our approach to implementation of TQM has not resulted in more paperwork, more committees, more meetings, and, in general, more bureaucracy.	5	4	3	2	1
Subtotal				______	
Total				______	

How to Calculate Your Score

To calculate your overall score, add the scores from the five categories and add the five scores together. For example, if you strongly agreed with all 40 statements and circled 5 for each one, your score would be 200.

How to Interpret Results

Use the following scale for interpreting your score:

200–160 Your organization's TQM effort is outstanding, and there is very little chance that the effort will be abandoned. Probably fewer than 5 percent of American companies are in this range.

159–130	Your organization has an advanced TQM approach, and it is likely that the effort will continue.
129–90	There is evidence of some commitment. Either your organization is in the early phases of the implementation or is feeling the presence of threatening circumstances. These threats probably will provide the motivation to continue with the effort.
89–50	Commitment is beginning to take hold in some parts of the organization, but there are few threats to motivate the effort. The chances your organization will drop the TQM effort within the next few years are greater than 50 percent.
49–0	Your organization's effort is doomed to failure, unless: major changes are made in the approach, commitment is obtained from more senior managers, or your business situation changes.

How to Use the Results

It is important not to put undue credence on the results of this survey. It is not designed as a scientific instrument, and it has not been validated. Neither is it based on empirical research that identified the characteristics of companies most likely to be successful with TQM. This survey is based on our experience of consulting with many successful and not so successful companies that have attempted to implement TQM. The major purpose of a survey such as this is awareness. It is intended to make the reader aware of the factors that impact the likelihood of success with TQM and that provide an indication of where an organization stands in relation to these factors. It is possible an organization could receive a low score on this survey and still have long-term success with the implementation of TQM. For example, Federal Express has never been in financial trouble, and it did not implement TQM to get its business back on track. It simply built its business on these principles and made customer satisfaction the number one priority for all employees. A strong score in the *commitment* area might make up for a lack of threat of any kind.

A low score in four of the five factors indicates that changes should be made in these areas. For example, if commitment to TQM is not strong, this should be changed. If executives are the problem, encourage them to talk to other executives in companies that have made TQM a way of life.

Similarly, the thoroughness of the plan can probably be improved, and so can your strategy for implementing TQM.

HOW TO APPROACH A TQM IMPLEMENTATION

As previously pointed out, several factors affect the success of a TQM implementation. Equally critical is the way these factors are coordinated to achieve the desired effect. The following discussion will explore two common approaches to a TQM implementation—the standard and the fast-track approaches—and will cite the advantages and disadvantages of each.

Standard Approach

Many organizations approach the implementation of Phase 1 of TQM by training employees and establishing quality improvement teams. Over the next several years, the Phase 1 effort is expanded to include more detailed training, such as process re-engineering and benchmarking. Quality improvement teams also are deployed across all functions, levels, and locations in the company. In many large organizations, this approach takes five years or longer.

As Phase 1 nears complete deployment, the organization starts to consider ways to further expand the impact of TQM. Usually, Phase 2 begins at this point, with an overall assessment of the organization's systems, processes, and results. Assessment findings are translated into plans to design new systems and re-engineer existing processes and systems. It often takes a minimum of two to four years for Phase 2 activities to impact the organization.

Seven to 10 years after beginning Phase 1, a company using the standard TQM implementation approach moves into Phase 3. In this phase, such total quality management concepts as prevention and empowerment are integrated into day-to-day operations.

By some estimates, in a large corporation the standard approach to implementing TQM takes one year for every layer of management. Smaller companies require less time because there are typically fewer levels of management. However, many small organizations still spend four or five years using this approach.

Fast-Track Approach

The fast-track approach is for organizations that do not want to spend 7 to 10 years to implement TQM. This approach is initiated with an assessment of the organization's systems, processes, and results. This assessment is the basis for the development of an implementation plan for Phases 1, 2, and 3. For example, Air Products & Chemicals in Allentown, Pennsylvania, is one company using the fast-track approach with a great deal of success. The Chemicals Group, with annual sales of over a billion dollars, has embarked on a multifaceted approach to fully deploy total quality management across the entire organization. The group uses assessments against the Baldrige Award criteria to drive the improvement effort—and develops aggressive plans to fill any holes uncovered in the assessment.

With the fast-track approach, Phases 1, 2, and 3 are not implemented in sequence, and there is a great deal of overlap. For example, while the training and quality improvement teams characteristic of Phase 1 are going on, systems in the organization are being re-engineered to reflect a more preventive approach to achieving quality and customer satisfaction. Similarly, some of the more sophisticated management approaches that are part of Phase 3, such as self-directed work teams, are started in the second or third year of the implementation. This fast-track approach is more risky than the standard approach. Some systems may fail because employees are not gradually accustomed to the changes. Changes are rapid and often significant, increasing the levels of stress in many employees. However, some organizations seem to thrive on rapid change and are very good at adapting to it.

The fast-track approach also presents more risk because it requires that major investments be made over a 3- to 5-year time frame, rather than the 10-year period common for the standard approach. As a result, companies short on cash reserves may find it difficult to adopt this approach. Ironically, the companies most likely to choose the fast-track approach are those that are in the most trouble and that often have the least resources.

With the fast-track approach, it will take a large organization about five years to reach the point where TQM is becoming completely integrated into the organization. Small companies that use this approach should expect to spend at least three years.

HOW TO MOTIVATE EMPLOYEES

Communicate Real Threats

It is much easier in the face of a crisis for an organization to secure commitment for a change effort like TQM. When employees realize that the company is in serious trouble, they rally to make changes and help the organization become healthy again. Executives in successful companies realize this, and often ask if it is necessary, or possible, to "manufacture" a crisis when one doesn't exist. The answer is both yes and no. An organization cannot tell employees that the company is in trouble when it is not or pretend there is a crisis when one does not exist. However, an organization can communicate real threats to all employees to help convince them of the need for change. For example, Air Products & Chemicals, a successful $3 billion company, is among the leaders in most markets it services. Air Products communicates the threats from competitors to all employees. The company explains that when you're in a leadership position, everyone is after you; and it shows employees how competitors are attempting to erode some of its market share by focusing on service quality. This helps emphasize the importance of not only maintaining product quality but of improving the service side of the business as well. Air Products' approach is not intended to instill fear in employees but, rather, to motivate them to change. The objective is to eliminate the sense of complacency often existing in successful companies that lead in their markets.

Communicate the Benefits of TQM

It is counterproductive to attempt to motivate people with an approach based solely on avoiding long-term negative consequences. The existence of a real or imagined threat may cause a change in behavior—but only until the threat is removed. When threats are used to motivate employees and managers to change, they may strive for improvement until the company is out of trouble, and then relax. Positive reinforcement is a far more effective way to motivate people and generate their interest in transforming the organization.

One of the most powerful ways to create interest in TQM is to communicate success stories. Moore Business Forms and a number of other companies hold annual sharing rallies where teams present results to each other in conferences of two or three days at a central location. At Moore, close to

500 employees are flown to Chicago for the three-day rallies. While it is a major expense for the company, the return on investment is that employees and managers at all levels become much more willing to take risks and implement changes to continuously improve the company's performance.

Another company that uses a similar strategy for sharing success stories is Cargill. Privately held, Cargill is not only one of the largest agricultural suppliers in the world, it is one of the world's largest organizations, with over $50 billion in annual sales and 70,000 employees across the world. For four years in a row, Cargill has held a Quality Conference in its Minneapolis headquarters to communicate the benefits of TQM and allow employees to share performance improvement ideas with one another. The 1993 conference, held in October, was attended by over 500 employees from all corners of the globe.

At the conference, quality awards based on the Baldrige criteria are given out to the best business units, and organizations get a chance to share their success stories to others. Like Air Products and some of the other companies we've discussed in this chapter, Cargill is number one in most of the markets it serves. Hence, there is no sense of threat that motivates employees to implement TQM. Like the others, Cargill uses a positive approach to communicate the benefits of TQM through the annual Quality Conference and other methods.

Another company that credits its current success to sharing business information with employees is Springfield Remanufacturing Corporation. Formerly, the company was an International Harvester plant. In 1983, when the company decided to close the plant, plant manager Jack Stack united employees and developed a plan to purchase the plant. The group raised $100,000 to buy the $9 million business. As Stack explains:

> We had an 89 to 1 debt to equity ratio, which put us on a par with, say, the government of Poland. [1]

In the 10 years since SRC was formed, the company went from a loss of $60,488 in the first year to $70 million in sales in 1993. The plant that started with 119 people now has 650 employees. The key to its success was teaching everyone the "great game of business," according to Stack. He discovered that it was not enough that every employee owned a piece of the business—everyone had to learn how to run the business as well. When his initial attempts to communicate financial information to employees were met with boredom and disinterest, he made business a game. Employees responded with interest, and Stack has spent the last several years teaching every employee to understand and play the "great game of business."

CONCLUSION

Two or three years from now, many of the organizations that today are working on TQM will be looking for a new consultant or a new program that promises to produce better results with less work—the easy way out. However, there are many more reasons to persevere with a TQM effort than there are to let it fade away. Hundreds of American companies have made major changes in their management practices and have achieved corresponding results with the TQM approach. In this chapter we examined some of the factors that will help increase the chances that your organization will be another one of those that achieves success. We also discussed two implementation strategies—the standard approach and the fast-track approach—that many companies use to pace the implementation to avoid the "too much too soon" syndrome and the "too little too late" phenomenon.

There is enough research to convince almost anyone that TQM really works. The key is to keep at it and to keep refining your approach through evaluation and improvement cycles. For organizations serious about TQM, hard work will pay off.

Tools, Tips, and Techniques
LAUNCHING A TQM EFFORT

The rationale for beginning a TQM effort is critical. Here is a list of do's and don'ts.

Don't	*Do*
Implement TQM to win an award or because you're being forced to by customers.	Implement TQM because it will improve the performance of your organization.
Wait until your business gets into trouble to begin TQM.	Begin TQM as a way of maintaining your current success and for ensuring success in the future.

Don't	*Do*
Try to implement TQM too quickly if your organization is not in a crisis mode.	Plan and schedule the implementation based on availability of resources and the urgency of the need to change.
Try to create artificial threats to motivate employees to change through fear.	Communicate real threats and the positive benefits of TQM as a way of encouraging employee support.

SUGGESTED READING

Bensen, Tracy E. "TQM—A Child Takes a First Few Faltering Steps." *Industry Week,* April 5, 1993.

Hammer, Michael, and James Champy. *Re-engineering the Corporation: A Manifesto for Business Revolution.* New York: Harper Business, 1993.

Stack, Jack. *The Great Game of Business: The Only Sensible Way to Run a Company.* New York: Currency Books, 1992.

Steward, Thomas A. "Re-Engineering—The Hot New Managing Tool." *Fortune,* August 23, 1993.

Chapter Three

Education

How to Get the Most from Your Training Investment

If you think training is expensive, you're right. Effective training is one of the most expensive change strategies in which you can invest. But training that fails costs even more.

In this chapter, we'll look at the four stages of quality related training and identify many of the reasons why it fails. We'll explore the reasons why employees are skeptical, why they enjoy the courses but don't learn anything, and why they learn new skills but forget them before they have a chance to apply them on the job. We'll also discuss why many organizations provide training in the wrong areas, and why managers who are taught new managerial techniques fail to change their traditional behavior.

We'll conclude the chapter with solid suggestions to help you design quality related training that should bring an excellent return on your training investment and actually yield improved bottom-line results.

WHAT KIND OF TRAINING—AND HOW MUCH?

After addressing the reasons for the pacing of the implementation, training then becomes a priority. A joint study conducted by Developmental Dimensions International and the Quality and Productivity Management Association revealed that training is the most important factor in a successful implementation of total quality management. [1] The research confirmed what most organizations already realize: training is an integral and essential part of the TQM initiative. Executives and managers know that employees must be trained if they are to understand the concept of quality and to master the use of the tools. However, given the diversity of content and the variety of delivery methods, many don't know what kind of training is appropriate, how much of it to provide, or to whom to provide it. The following discussion will offer insights into this critical aspect of the TQM implementation.

What Kind of Training Is Required

Organizations that achieve success in quality related training efforts typically provide training in four stages. The first, **conceptual training,** exposes employees to the concept of quality and the impact it can have on an organization. **Quality tools training,** the second stage, provides the basic tools and training on how to use them. The third stage of training, **special topics,** addresses topics and aspects of quality specific to job function, and the fourth, **leadership training,** is aimed at executives and managers who lead the initiative. A successful TQM implementation requires an appropriate balance of all four stages of training across the organization.

Quality concept training. Quality concept training has evolved from the philosophical training of the early 1980s. The earliest training was based on the philosophies of three of the movement's best known gurus—W. Edwards Deming, Philip Crosby, and Joseph Juran. Its objective was to teach employees to think differently about quality and to effectively manage their organization in a new environment. Some of the most popular courses were offered by Crosby's Quality College in Florida and by Deming's and Juran's workshops.

Similarly, today's training is based on philosophy, but it is the philosophies and concepts of the organization that are taught. Quality concept courses are not intended to teach specific skills or to change employee behavior on the job. They are designed to lay the foundation for the subsequent quality tools training, and the measure of success is the amount of enthusiasm they generate.

Quality tools training. In this stage of training, employees are taught to channel that enthusiasm into actions. Quality tools training provides employees with an understanding of the quality tools and how to apply them in their day-to-day work environment. In this stage of training, employees are equipped with the knowledge and skills to analyze problems and improve their performance. Typically, employees learn how to draw Pareto charts, how to prepare cause-effect diagrams, and how to create a control chart. These basic quality tools often are taught in a fun workshop, using such props as blocks, beads, or small parts that need to be soldered.

Special topics training. Most large corporations and many smaller organizations already have trained employees on the basic concepts and tools of total quality management. Now, in the 1990s, training dollars are being spent on special topics courses that address such areas as:

- ISO 9000.
- Malcolm Baldrige Award criteria.
- Quality function deployment.
- Measuring customer satisfaction.
- Design of experiments.

Leadership training. Leadership training has evolved from what used to be called "management development" or "supervisory training." To accommodate the shift in management style that TQM requires, new courses have been put together to teach managers how to inspire and empower employees and how to break down barriers that keep employees from reaching their goals.

How Much Training Is Necessary

Some organizations are frugal with training resources and allot minimal time to the training experience. These organizations try to pack everything into one eight-hour course. Often they discover that employees are able to review the content in eight hours, but they don't learn the concepts and usually don't acquire the skills or the capability to apply them properly.

At the other end of the scale are companies, especially the Baldrige Award winners, that claim to provide employees with as much as 80 hours of quality related training, in addition to the training in leadership skills provided for managers. While it is difficult to determine precisely how much training is too little, and how much is enough, observation of organizations that are fairly far along in the implementation of TQM indicates that about 40 hours of training is a minimum for the first year. This training should be apportioned as follows:

- Quality concept training: 8–12 hours.
- Quality tools training: 16–24 hours.
- Special topics training: 4–16 hours.

After the first year of the quality implementation, about 16–32 hours of additional training on quality topics should be provided each year.

In addition to the training requirements in areas of quality concept, quality tools, and special topics, during the first year managers should receive an additional three to five days of training on how to manage employees in a TQM environment. Managers also will need follow-up leadership training in subsequent years. As the organization moves into Phases 2 and 3 of the quality journey, managers need training on such skills as coaching, empowerment, and working in a self-directed team organization.

Who Needs Training

Every employee in the organization needs the minimum amount of training in areas of quality concept and quality tools. Each major group of employees also requires special tools training specific to the group's functional responsibilities. Executives and managers who lead the TQM effort need more training than any other level of employee. This training should be in the three areas of quality concept, quality tools, and special tools, as well as in leadership skills required to guide the initiative.

WHY TRAINING FAILS

Many of the reasons that quality related training fails—and training dollars are wasted—have more to do with the organization's approach to training, and the way it is delivered, than with the concept of training itself.

In most organizations, the objectives of training are threefold: (1) to accustom employees to a certain mode of behavior and level of performance, (2) to establish proficiency through instruction and practice, and (3) to motivate employees to make desired behavior changes. In many organizations, training is approached in a manner that makes it difficult to achieve these objectives—if not impossible—and this is when training fails.

The following discussion addresses many of the reasons why training, even training implemented with the best of intentions, often fails.

Failure in Quality Concept Training

Many of the quality concept courses taught in the early 1980s failed to achieve their objectives because participants were not convinced of the value of the new approach or of the company's willingness to pursue it.

In quality concept training, it is common for employees to revert to skepticism—often the message and the approach are so contrary to the way the organization operates that people simply can't believe the company can change enough to adopt it. For example, one organization spent more than $1 million to send employees to Phil Crosby's Quality College in Florida and to conduct follow-up courses with Crosby's materials back home. When the training was completed, there were still a number of skeptics in the organization. In fact, in one group of people, only one in four was convinced the new approach was destined to become the way of the future. The attitudes of the remainder ranged from somewhat convinced to outright rejection.

Several other reasons why quality training fails in the concept phase are listed below. Let's discuss them one by one.

Unrealistic expectations. Training often is ineffective if organizations have unrealistic expectations in the early stages. Quality concept courses are not intended to improve quality by themselves. The goal of the training is to convert employees to a new way of thinking about quality. Despite this, many expect to experience an improvement in quality immediately after the course. At the organization in the previous example, product quality levels actually declined during the training period. The company was outraged, until Crosby pointed out that the company had not changed any of the other 15 or 20 things that he identified as higher-priority items.

Training not tailored to the audience. Quality concept training often fails because the training is not tailored to the audience. Typically, courses are packaged training, designed with a "one size fits all" approach. Everyone from the CEO to the mail room workers attend the same courses. Individuals in big companies, small companies, manufacturing, and service companies all receive the same training.

Training is irrelevant. Employees in support functions and service organizations frequently believe that quality concept training does not apply to them. In the 1980s, most of the training programs on the market were designed by and for individuals with backgrounds in big manufacturing companies. Service employees view manufacturing related examples and exercises as irrelevant. They agree that concepts like achieving quality through prevention of errors, rather than detection, makes sense for plant and engineering people, but they argue that it doesn't apply to people in support functions or service organizations.

Employees doubt the company's commitment. Many organizations, even today, use a defect-detection approach to quality, rather than the preventive approach espoused by quality experts like Deming, Crosby, and Juran. Management priorities are short-term financial results, not quality or customer satisfaction. Employees who attend the training agree that the new approach to quality makes a great deal of sense. However, they don't believe the company is serious about the effort. They have seen other programs and initiatives come and go over the years and, with good reason, believe TQM is just another management fad.

Training doesn't start at the top. The approach that organizations use to roll out the training often is responsible for its failure on two counts. Since training is typically directed at middle management and employees at lower levels, executives do not receive the training they need to lead the effort. This practice also tells employees the company is not serious about the effort and leads them to question the level of commitment at the top.

Lack of follow up. In many organizations, employees complete the training and hear nothing more about TQM for a year or longer. Things may be happening behind closed doors, but most employees don't realize it. Consequently, they think the effort has died. In other organizations there is follow up, but the actions amount to little more than hype. There is considerable fanfare—quality buttons, posters, banners, and inspirational speeches from executives—but nothing concrete happens.

Failure in Quality Tools Training

Organizations spend significant amounts of time and money to develop and deliver quality tools training. Many develop their own courses while other purchase packaged training materials. Either way there are inherent shortcomings that result in ineffective training.

Many large manufacturing companies favor a highly interactive workshop, where participants are exposed to a variety of quality improvement tools but are not provided with the practice necessary to master them. Employees who complete the course recognize the various tools and know how they are used. Often, though, they can't apply the tools to real problem situations in their work environment. This illustrates the main reasons why quality tools training is often ineffective: participants do not get enough practice with the tools to master their use, and they do not get enough training to apply the tools in their own work situation.

Not enough practice. Studies indicate that the most direct path for adult learning is a three-step process of lecture, demonstration, and practice. Many popular courses are structured to provide about an hour of lecture and demonstration on each tool. Participants are told what the tool is and are shown what it looks like and how to use it. However, in most workshops, there is no opportunity for participants to use the tool themselves.

In courses that do include skill practice, participants are required to perform exercises using each of the tools. Although better than no skill practice, this often is not enough. Some of the simpler tools can be learned with a single exercise, especially if the audience consists of engineers or other professionals who are familiar with graphs, statistics, and data. Other tools are more complex and require repeated practice to master them.

Lack of applicability. There are quality tools courses available that provide participants with sufficient practice to master the use of the tools. However, these courses do not provide participants with an understanding of how to apply the tools in their own work environment. Back on the job, employees cannot translate what they learned in the course into an overall approach they can use to improve their own processes and performance.

Failure in Special Topics Training

Most organizations today are in the special topics stage of training, and this is where the major portion of the training budget is spent. Many large corporations have developed very thorough catalogs of quality related courses. Many of these courses, when examined on an individual level, are excellent. However, a look at the whole catalog reveals the lack of a systematic curriculum that specifies sequences, prerequisites, and suggested courses for particular functions and levels of employees. This inconsistency is the root cause of most training failures in this special topics stage of training.

Systematic needs analyses not conducted. Courses often are purchased from vendors, or developed in-house, without an overall plan. In many organizations, training needs analysis consists of allowing employees to identify their interests from a list of topics. This wish list approach is little more than a good way to identify the training topics in which employees are interested. It rarely results in a collection of courses that adequately fit the organization's true training needs.

Courses not logically aligned. Many organizations don't consider curriculum design until they've amassed a large collection of courses. At that point, force-fitting a curriculum architecture on a collection of existing courses is like trying to design a car when the individual systems already have been built. Yet, this is the way many organizations approach curriculum design, particularly in the area of special topics courses.

Failure in Quality Leadership Training

Total quality management is much more than a collection of new tools and techniques for measuring and improving quality. TQM requires a new way of managing. In the book *Out of Crisis,* W. Edwards Deming explains:

> The job of management is not supervision, but leadership. Management must work on sources of improvement, the intent of quality of product and of service, and on the translation of the intent into the design and actual product. The required transformation of Western style of management requires that managers be leaders. [2]

Many managers and executives have struggled with this new approach to leadership. The familiar management techniques of directing and controlling employee behavior do not elicit the desired results within the TQM framework. As a result, managers, many who are in their 40s and 50s, are being asked to make dramatic changes in the way they manage their organizations.

Organizations do not expect managers to make these massive behavioral changes without help. Most expend considerable resources on new training courses intended to teach managers how to lead, empower, and inspire their employees. Most of today's leadership training courses, however, are nothing more than yesterday's ineffective management training programs, renamed and revised to reflect a new approach.

These programs focus on exercises instead of lecture, and the presentation often takes advantage of recent advances in multimedia technology, but the mechanisms that led to failure in the earlier programs are still inherent.

Almost 25 years ago, a classic article was published in the *Training and Development Journal,* comparing management development training to the emperor's new clothes. The authors concluded that, even though everyone wanted to believe most management development programs changed manager's behavior, very little evidence indicated that they did. Similarly, the ability of today's programs, like the outdated programs they've replaced, have a questionable ability to change managerial behavior.

There are four major reasons why the TQM leadership courses currently being used in organizations usually fail. We'll discuss each of them.

Antiquated theories. Leadership courses often fail because too much time is spent teaching managers antiquated theories of motivation. Almost 30 years ago, George Odiorne, father of management by objectives, or MBO, warned of the futility of teaching managers motivation theories:

> Has all this talk and work on teaching motivation theory been overdone? I'd like to suggest that it has. Training should change job behavior. It has no useful purpose in teaching managers to probe in private motives. [3]

Despite this, the theories of Hertzberg, Maslow, and McGregor that managers were taught 30 years ago are still being taught today. Even though these motivational theories are valid, it is doubtful they have a significant impact on the ability to manage. They are included in leadership courses in the belief that insight into factors that motivate employees makes better managers. While it might be beneficial if managers were given the extensive amount of motivation theory training that psychologists receive, the limited training managers receive is not enough to provide the perspective and depth necessary for complete understanding.

Pop psychology models. Pop psychology management models and matrices gained popularity in the 1960s and continue to be used in the 1990s. They are, essentially, a simple way to classify and label employees. The underlying logic is that, when managers learn which boxes individual employees fit into, they can adapt their management style to fit the individual. Some of the early models taught to managers were:

- Lewin, Lippitt, and White's three styles of leadership: democratic, laissez-faire, and autocratic.
- McGregor's styles of management: Theory X and Theory Y.
- Blake and Mouton's managerial grid: 1/1, 1/9, 5/5, and so on.
- Hershey and Blanchard's situational leadership model: directing, coaching, supporting, and delegating.

Some of these classification systems are being taught in today's leadership courses. Phil Crosby has recently come out with his own grid for evaluating an organization's leadership style. In his book, *Completeness* (1992), Crosby presents a matrix that classifies organizations into five styles of leadership [4]:

- Third Reich.
- Banana Republic.
- Constitutional Monarchy.
- American Republic.
- 21st Century Completeness.

These matrices and models are popular because they reduce a complex universe to a simple world, and therein lies their failing—they are too simplistic. People and organizations do not fit neatly into boxes or squares on a grid. Just as it is unlikely that learning motivation theory will make a better manager, there is little evidence to suggest that a person who has learned to classify employees as a 1/9, an "analytical driver," or a "transformational leader" will be a better manager.

A second factor casts additional doubt on the effectiveness of these models and matrices—they are not good predictors of behavior. Labeling an employee according to these psychological models does not increase the manager's ability to predict or control their behavior.

Self-discovery tests. Leadership training programs have a third common characteristic that often leads to failure. Typically, these programs include a test or exercise that participants complete to gain insight into their own personalities. A commonly used evaluation is the Myers-Briggs, which classifies people into 16 different categories or boxes, based on the following dimensions:

- Extroversion versus introversion.
- Intuition versus sensation.
- Thinking versus feeling.
- Judging versus perceiving.

Exercises like the Myers-Briggs often have a great deal of face validity. People are seldom surprised when they see how they are classified, and they generally believe their personality is described accurately. However, this insight alone has little effect on a manager's ability to be a better leader.

The greater value of self-discovery tests is that they are fun and a good way to inject a lighter note into otherwise long and intense sessions.

Lack of applicability. Lack of follow-up is the biggest reason why the leadership training programs in many organizations fail to change behavior. Learning new skills in a workshop is one thing. Applying them on the job is much more difficult. This is especially true when managers learn new skills that may be contrary to their own style of managing. Most organizations fail to provide the feedback and follow-up that managers need to determine how successfully they are implementing the new management techniques. As a result, only a small percentage of the managers who attend the leadership courses actually change the way they manage.

INCREASING THE CHANCES FOR TRAINING SUCCESS

Organizations can take several steps to help ensure successful experiences in each of the four stages of training. We'll address each of the factors in the following discussion.

How to Make Quality Concept Training More Effective

Use relevant examples. One important way to enhance the value of quality concept training is to design examples that are relevant to employees in the organization. Ideally, examples should reflect actual situations that employees encounter in their jobs. When this is not possible, examples and exercises should be based on experiences in similar companies.

Tailor training to the organization. Training should be customized to the organization's needs and accurately reflect its culture. Packaged programs can be tailored to suit your needs, and most training vendors offer this service. Programs also can be customized in-house, by rewriting exercises and examples to more closely reflect the organization. In most cases, it will probably be more cost-efficient to purchase a packaged program than to develop one from scratch.

Implement training at the top. Training should be delivered top-down, starting with senior executives and cascading down to the lower levels in the organization. Executives should be required to take the full session, rather than an abridged version of the training. It also is beneficial if executives receive additional training focused on the roles and responsibilities they have in leading the company's journey toward total quality management.

Follow up with concrete actions. The effectiveness of quality concept training is greatly enhanced when it is followed up immediately with specific plans and changes in organizational systems that focus on the prevention of problems and on continually delighting customers. For example, an organization might announce that it is embarking on a new comprehensive approach to measure customer satisfaction as part of the quality effort. Also, at each step of the implementation, employees should be kept aware of what is happening.

How to Make Quality Tools Training More Effective

Provide time and opportunity to master skills. This is the most significant action an organization can take to improve the effectiveness of training. In quality tools training courses, employees should be given sufficient time to practice the skills they are being taught. The most effective training incorporates the exercises that give employees an opportunity for hands-on experimentation with each tool.

Provide a framework to use skills in the work environment. Successful quality tools training requires that participants acquire an understanding of how to apply the tools in their own work environment. Ideally, the training should be spaced over time to lead intact teams through an entire quality improvement process. One series of courses developed by the Council for Continuous Improvement (CCI) was recently redesigned to include training on how to use the tools to improve work processes. CCI is a consortium of companies that have pooled knowledge and resources to develop quality related training materials that meet the needs of all the companies in the consortium. CCI offers a series of about 15 individual two-hour training modules on quality improvement tools, including:

- Process flow.
- Cause and effect.
- Data collection.
- Pareto charts.
- Histograms.
- Graphs.
- X & R control charts.
- X & S control charts.
- X & MR charts.
- np control charts.
- p control charts.
- c control charts
- u control charts.
- Using and interpreting control charts.

The modules are well designed from an instructional standpoint, and they include many opportunities for participants to practice with the tools and apply them to problem situations.

How to Make Special Topics Training More Effective

Train employees in the right areas. Special topics training should reflect the organization's strategic business goals—and business and quality improvement goals should be the driver of training needs. When training goals have been aligned with the overall business goals, individual functions and jobs should be evaluated to determine which employees need training, and in which topics they need to be trained.

Organize courses into a logical curriculum. Organizations that have acquired a library of several training courses need to organize the courses into a systematic curriculum that specifies sequences, prerequisites, and suggested courses for particular functions and levels of employees. Many organizations that have adopted the criteria of the Malcolm Baldrige National Quality Award use the Baldrige criteria to sort courses and provide some structure to a varied collection of individual modules.

There are two points to keep in mind when organizing courses into a curriculum: (1) sort courses by level, with beginning, intermediate, and advanced levels grouped together; and (2) determine which of the foundation courses are required for all employees and which can be classified as "electives." A typical quality training curriculum is shown in Figure 3–1.

Develop suggested training paths for each major group of employees. Special topics training also is more successful when training paths are developed for such functions as engineering, finance, production, operations, and others. The training paths should be tailored to the needs of the jobs in those functions. General Dynamics' Fort Worth Division is one of the few companies that designed its quality curriculum in this manner. Suggested training paths and required and optional courses are identified for various functions of employees. For example, all employees might be required to complete the four basic courses that outline the company's quality philosophy and approach. After these core courses are completed, separate training paths are identified for particular job functions, with both required and optional courses.

Establish curriculum design early. Designing a curriculum before courses have been purchased or developed allows the flexibility,

FIGURE 3–1
Generic Quality Training Curriculum

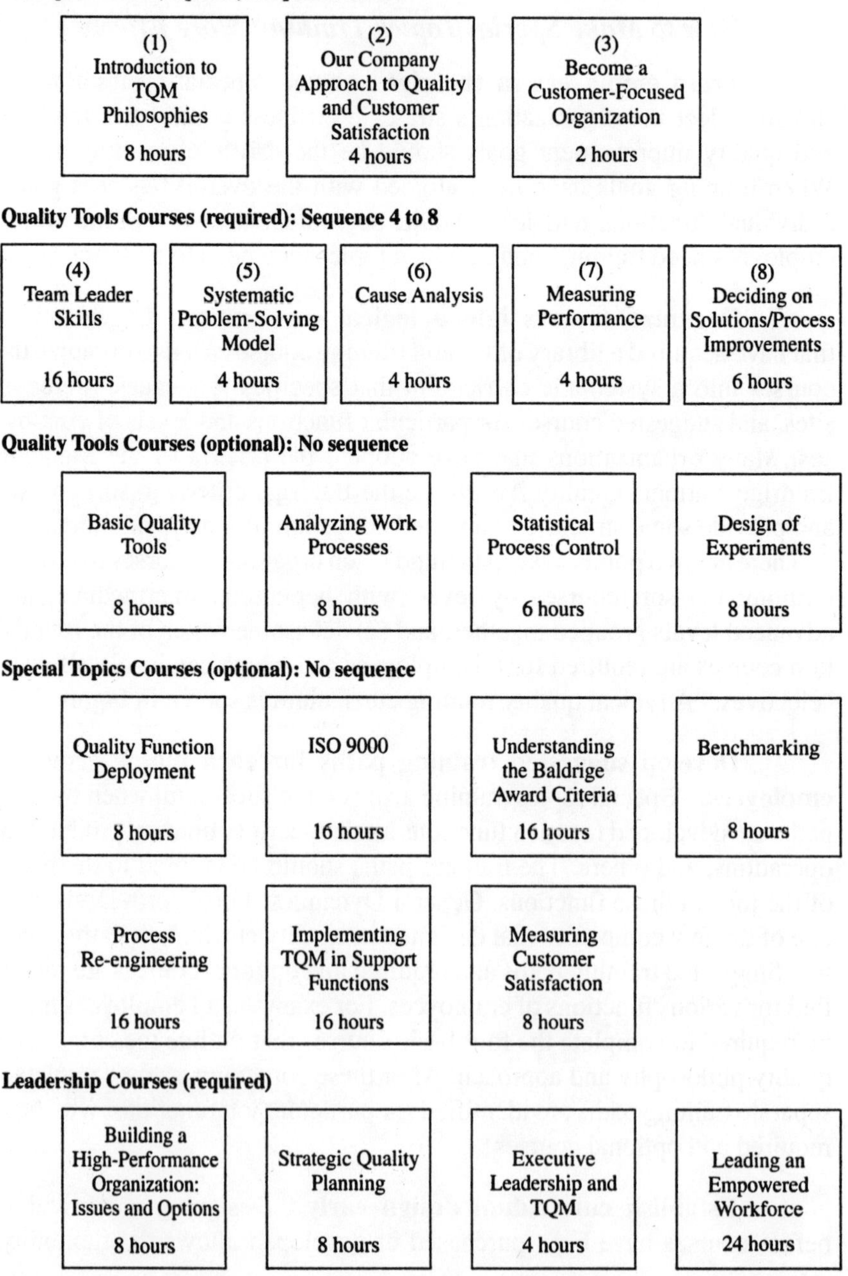
Concept Courses (required): Sequence 1 to 3
(1) Introduction to TQM Philosophies 8 hours
(2) Our Company Approach to Quality and Customer Satisfaction 4 hours
(3) Become a Customer-Focused Organization 2 hours
Quality Tools Courses (required): Sequence 4 to 8
(4) Team Leader Skills 16 hours
(5) Systematic Problem-Solving Model 4 hours
(6) Cause Analysis 4 hours
(7) Measuring Performance 4 hours
(8) Deciding on Solutions/Process Improvements 6 hours
Quality Tools Courses (optional): No sequence
Basic Quality Tools 8 hours
Analyzing Work Processes 8 hours
Statistical Process Control 6 hours
Design of Experiments 8 hours
Special Topics Courses (optional): No sequence
Quality Function Deployment 8 hours
ISO 9000 16 hours
Understanding the Baldrige Award Criteria 16 hours
Benchmarking 8 hours
Process Re-engineering 16 hours
Implementing TQM in Support Functions 16 hours
Measuring Customer Satisfaction 8 hours
Leadership Courses (required)
Building a High-Performance Organization: Issues and Options 8 hours
Strategic Quality Planning 8 hours
Executive Leadership and TQM 4 hours
Leading an Empowered Workforce 24 hours

and efficiency, of purchasing or creating only those courses that are needed and that have a logical place in the curriculum.

How to Make Leadership Training More Effective

Provide ongoing feedback. Managers who have completed leadership training need feedback to determine how successfully they are implementing the new management techniques. Northrop Corporation ensures that leadership training actually changes how managers manage through the use of feedback instruments both before and after the training.

Six weeks prior to the Northrop managers' leadership training, employees fill out an anonymous questionnaire on their bosses' managerial behavior. Data are tabulated and results, including specific areas that need improvement, are reviewed with the managers in the leadership course. About eight weeks after the managers have completed the workshop, their employees fill out the same survey again. Individual coaching sessions are scheduled with the workshop facilitator so each manager can receive individualized coaching on areas of strength and on areas needing improvement.

The management practices taught in the leadership course were derived from the company's six values, and these are integrated into performance planning and assessment as an additional way to reinforce behaviors that are consistent with the values.

Although, despite this approach, some managers still fail to change their management style, it greatly increases the chances that leadership training will be successful.

CONCLUSION

When training fails, often it is not the concept of training that's responsible. Usually, it is the content and method that lead to disappointing results. In this chapter we discussed the four stages of quality related training that organizations typically provide, along with the key factors that lead to failure in each stage. We also offered numerous suggestions that will help organizations develop effective training programs to meet their objectives. Training is still one of the best investments an organization can make, but only if it's done wisely.

Tips, Tools, and Techniques
SUCCESSFUL TECHNIQUES TO MAKE QUALITY TRAINING MORE EFFECTIVE

Develop exercises and examples that fit the audience, rather than using generic ones.

Deliver training in a just-in-time fashion right before employees will use the knowledge/skills.

Minimize the amount of lecture in all training—interaction is the key to learning.

Use performance and paper/pencil tests to evaluate participant's mastery of the material.

Train employees and supervisors together when they are from the same department.

Schedule a follow-up assignment after the training to ensure that skills are used on the job.

Develop individualized training paths for each major job function.

SUGGESTED READING

Barker, Julie. "Labor Pains . . . Management Training Is Suffering and the Problem Stems from the Top." *Successful Meetings,* November 1990, pp. 1–4.

Crosby, Philip B. *Completeness.* New York: New American Library/Dutton, 1992.

Deming, W. Edwards. *Out of Crisis.* Cambridge, Mass.: MIT Center for Advanced Engineering Study, 1982.

"The Emperor's Clothes." *Training and Development Journal.* Training Research Forum at Harrison House, June 1970.

Ginnodo, William L., and Richard Wellins. "Research Shows That TQM Is Alive and Well." *Tapping the Network Journal,* Winter 1992/1993, pp. 2–5.

Keirsey, David, and Marilyn Butes. *Please Understand Me: Character and Temperament Types.* Del Mar, CA: Prometheus Nemisis Books, 1978.

Svenson, R. A., and M. J. Riderer. *The Training and Development Strategic Plan Workbook.* Englewood Cliffs, NJ: Prentice Hall, 1992.

Chapter Four

Results

How to Get TQM to Pay for Itself

A total quality management implementation does not require several years of faith before results begin to materialize. An organization should expect results during the first year, and every year, of its TQM journey. Most of the Phase 1 strategies produce a return on investment within a short time of their implementation. However, many organizations do not see that return. It's usually because they focus their quality efforts on short-term cost-cutting endeavors or on activities that waste time, money, and other precious resources. Typically, these activities do not lead to the improved performance, the increased customer service, and the more efficient processes that drive bottom-line operational and financial gains. Rather, they promote the wrong measurements, reinforce the wrong behavior, and encourage the wrong attitude.

In this chapter, we'll address the reasons why organizations tend to put too much focus on activities and not enough on the results the activity should generate. We also will identify some of the common mistakes that result in wasted time and money. We'll conclude the chapter with suggestions that will help you balance activities and get the results to achieve the bottom-line improvements you anticipated.

WHY ORGANIZATIONS FOCUS ON ACTIVITIES INSTEAD OF RESULTS

It is common for organizations to overfocus on activities that are the means to an end and to disregard the end—the results. The tendency is a by-product of the total quality management movement. For years, quality gurus like Deming and Crosby admonished American management for the inclination to focus on short-term financial results in preference to spending resources on activities that would produce long-term gains in quality and customer satisfaction. In response, many organizations de-emphasized short-term financial results and increased their emphasis on factors that would lead to longer-range improvements. In the mid 1980s, for example, Canadian telecommunications giant Northern Telecom required every employee, from senior executives to individual contributors, to set at least one "improving the business" objective with a long-term focus. Since results would not be apparent for several years, Northern Telecom, like most other companies that set long-range goals, gauged intermediate success in achieving the objectives by looking at activity measures.

Until recently, the tendency to focus on activities over results also has been reinforced by two of the most credible frameworks for the implementation of TQM, the Baldrige Award criteria and the criteria for ISO 9000, both of which promoted this method of keeping score. In fact, roughly 650 out of 1,000 possible points in the Baldrige Award criteria are based on the activities and approaches used in the organization. Only 350 points are based on results.

The criteria asked for trend data on indexes, such as the number of teams that exist, the amount of money the company spends on quality related training, and the number of quality related awards and recognitions handed out. In the section dealing with human resource practices, the emphasis was particularly heavy.

In 1993, the Baldrige criteria were revised to shift emphasis from data requirements that encourage companies to measure activities to those that demonstrate the *effectiveness* of key activities. For example, instead of data on how many hours of quality related training that employees receive, companies are asked to provide data demonstrating that the training resulted in improved job performance.

Like the earlier Baldrige criteria, criteria for receiving ISO 9000 certification are based almost exclusively on a company's activities, encouraging such activities as documentation and control of key processes. No results are examined during ISO certification. Phil Crosby explains his views on ISO 9000 and the Baldrige Award criteria in the May issue of *The Quality Observer:*

> The governments have the illusion that they can make up rules about quality so they have ISO 9000 and the Baldrige criteria but they adopt the philosophy that if companies follow those, they will become world-class quality companies, but it's just not so. Europeans are trying to make other people as inefficient as they are, by using ISO 9000. [1]

COMMON MISTAKES THAT WASTE TIME AND MONEY

Many organizations invest in strategies that focus on the wrong outcomes. The activities they implement, or the manner in which they implement them, do not lead to improved performance, to increased customer service, and to more efficient processes. Rather, they emphasize the wrong indicators, reinforce the wrong behavior, and encourage the wrong attitude. We'll address many of these counterproductive strategies in the following discussion.

Measuring the Wrong Indicators

Many large manufacturing companies measure the effectiveness of their total quality management efforts by collecting data on such considerations as the following:

- The number of hours employees spend in training each year.
- The number of teams in place.
- The percentage of employees involved on teams.
- The percentage of employees who can recite the company's mission and values.
- The number of processes for which process models have been developed.
- The number of improvement suggestions submitted.
- The number of quality related awards given out.

Although these activities are easily measured, and data are easy to collect, this inclination to measure the activities instead of the results leads companies to invest in strategies that often are counterproductive. For example, a goal of one company, which spent five years in Phase 1 of total quality implementation, was to have 100 percent of employees on quality improvement teams. Several years were devoted to training and implementing the teams. By 1991, more than 100 percent of employees were on teams—some were on two or three quality improvement teams. However, during these five years, important measures of customer satisfaction, productivity, cycle time, financial, and operational results did not improve. In fact, the finance department reported that productivity per employee had gone down due to the amount of time employees spent in team meetings.

When a member of one team was asked why the team was having a meeting, the response was, "It's Friday at two o'clock; we always have team meetings on Fridays at two." When asked what the team was working on, team members explained that they hadn't yet settled on a particular problem or process. This company also found that, when quality improvement teams *did* define areas to address, many of them worked on issues that were important to them but often were not tied to improving the company's performance.

In 1992, the company reevaluated the emphasis on teams and is now phasing out many of them. The company's current goal is to have no more than 50 percent of its employees involved on teams, and teams are provided with more direction in selecting problems and issues on which to work.

Another company decided that a good measure of its success with TQM was the percentage of employees who had attended quality related courses. Objectives were written and achieved when all employees received the mandatory 24 hours of quality related training that was specified in the objectives. In the case of one woman, this training consisted of eight hours of problem-solving tools training, eight hours of team member training, and eight hours of team leader training.

Nine months after the training, this woman was asked how she had applied what she learned in these courses. She explained that she had forgotten most of it. Not only had she never led a team, she had never been asked to be on one!

Many companies measure progress with TQM by tracking the number of hours of quality related training they conduct. Believing that more hours in training automatically equated to higher quality of training, one company had a course that required employees to be in class eight hours a day for 26 weeks. Employees who took the course, which was mostly theory, certainly received more training. But whether they received better training is questionable, especially since the course included no testing measures.

Focusing on Behavior instead of Accomplishments

In many organizations, particularly service organizations, processes are controlled by human beings, rather than by machines. In others, in such functions as research and development or engineering, the accomplishments of employees are difficult to measure. As a result, many organizations put considerable emphasis on trying to control the behaviors of employees, rather than their accomplishments. The flaw inherent in focusing on behavior measures is that the means do not always lead to the right ends. Employees can be trained to behave, with regularity, in a certain manner, but the behavior often does not lead to a worthwhile accomplishment.

For example, the managers of a major supermarket chain wanted employees to appear friendlier, so they decided to have checkers smile and say, "Have a nice day" as they handed a customer the grocery receipt. The reward system created to encourage the behavior worked exceptionally well, and the "Have a nice day" program was considered to be successful. Baseline measures showed that clerks told customers to have a nice day an average of 7 percent of the time. Within 12 weeks, close to 85 percent of all transactions with customers were concluded with the clerk telling the customer to have a nice day.

The flaw in the program, however, was that this company spent a great deal of time and money encouraging checkers to say something 75 percent of their customers either didn't want to hear or didn't notice hearing. Subsequent focus groups and customer interviews revealed that more than 50 percent of the supermarket's customers did not know whether they'd been told to have a nice day. About 25 percent said hearing the phrase was all right, but the remaining 25 percent said they did not like hearing "Have a nice day" at all.

In his classic text, *Human Competence* (1979), Thomas Gilbert suggests that organizations focus too much on trying to control behaviors (B measures) of employees, rather than their accomplishments (A measures). Gilbert says:

> We have no need to measure behavior until we have measured accomplishments. Quantitative expressions of behavior (B measures), except for special purposes, are often misleading indices of performance. What we want as a result of measuring the behavioral side of performance is a list of deficiencies that are significant only because they lead to important accomplishment deficiencies. [2]

In the supermarket example above, even though it was possible to motivate employees to change behavior, the behavior did not increase the likelihood that customers would buy more or shop at that store in the future. It was not an important behavior change and had little, if any, impact on improving bottom-line results.

Emphasizing Courtesy instead of Competence

In many organizations, customer contact employees are trained to smile, chat with customers, and make them feel better when a problem occurs. As a result, it is common to encounter poor quality service that is delivered by incompetent employees who are extremely pleasant and courteous.

One reason so many organizations, even organizations implementing TQM, focus on courtesy, rather than on competence, is because courtesy is inexpensive and easy to achieve. Most people can be trained to be courteous in a good one-day workshop. It takes a great deal more time and money to teach employees to be knowledgeable and competent in their jobs, and it often requires hiring employees who are more skilled.

Employees who are courteous, friendly, and pleasant to deal with do not compensate for incompetent performance. Chris Lee of *Training Magazine* says:

> Given a chance, I'll deal with people who are friendly and competent. But, if I only get one out of two, I'll take competence every time. [3]

Organizations that focus training efforts on courtesy, and neglect training employees to competently perform their jobs, usually are among those that fail to see bottom-line improvements.

Disguising Cost Control as TQM

Many organizations present cost-cutting programs in the guise of total quality management, particularly when it's doubtful employees would support a cost-cutting effort. Typically, these companies train all employees on the basic quality tools, then form teams and start the teams working on improvement projects. However, the teams are allowed to work only on projects that will save the company money in the short term. Efforts to improve softer measures, such as quality or customer satisfaction, are not considered to be a good use of team time.

One restaurant chain that adopted this approach to TQM did not have good measures or data on levels of quality or customer satisfaction. However, the organization did have data on a major performance index, the amount of chicken that was sold versus the amount thrown away—in other words, scrap rate. All the projects the organization's many teams pursued were directed at improving this index or at improving other financial measures, such as labor costs or overhead expense.

This organization and a number of other organizations that follow the same path overlook the underlying premise of TQM—that continually delighting customers is rewarded with long-term success. Concentrating on quarterly financial results and cost cutting are short-sighted strategies that, by themselves, will not ensure customer satisfaction or long-term success.

Focusing on Internal instead of External Improvements

Too much internal focus is another reason many organizations fail to achieve bottom-line results. It is common for these organizations to spend months working to improve a process or an aspect of performance that is of little or no importance to internal or external customers. Dr. John Call of Gilroy Foods tells a story about a quality improvement team in his company that spent a year working to reduce the cycle time required to process a customer's order.

By changing the process and simplifying procedures, the team was able to significantly reduce the time between placement of the customer order and delivery of the order to the customer's dock. However, when customers were presented with the results of the project, they explained that the company had done fine on order processing time before the project. They were concerned about product consistency, not cycle time.

In a major corporation, the human resources function that designs performance appraisal systems also neglected to ask customers what was most important before beginning an improvement effort. The team embarked on a two-year project to design and implement a world-class performance appraisal system. The new system they designed and implemented met every one of the relatively complex objectives and even received an award from a human resources professional society.

Despite these positive factors, the new system was not well received among managers and employees who had to work with it. They objected to the eight-page form (the old one was three pages), the requirement for quarterly reviews (the old system required annual reviews), and the fact that administering the performance appraisal system required a manager to spend about 40 hours per year for each subordinate.

Time and money are being wasted on internally focused process improvement activities in organizations all over America. Many of these activities are doing nothing to increase customer satisfaction, and some, like the human resources example above, are actually alienating customers. In many instances, the simple step of identifying customer priorities and requirements would have resulted, instead, in improvements that customers welcomed.

Failing to Identify Key Process Variables

One of the major tenets of the total quality management philosophy is process control. This premise has led many companies to concentrate on measuring and controlling their processes to improve their accomplishments or results. In many cases, the approach has been quite successful. By analyzing processes, improving them, and controlling key variables, results have been more consistent and quality has improved.

Many other companies, however, have not achieved this success. The reason is that achieving quality by measuring and controlling process variables is effective only when both the variables and the levels of performance that lead to desired levels of product/service quality have been empirically identified. It is common to assume that certain process variables will produce the desired results. In organizations that fail to do the thorough research necessary to identify the cause-effect relationships between behaviors or process variables and results or outcomes, attempts at process management result in wasted time and money.

Ineffective Benchmarking

Benchmarking, in its truest sense, can be a valuable activity, but when used inappropriately it is little more than a corporate field trip. For example, many managers initiate trips to other companies, particularly those that have won the Baldrige Award, to observe their quality processes. These trips are usually too broad in focus to be considered benchmarking, and often the individuals have prepared neither a plan for the trip nor a list of data to gather.

Benchmarking also can be a waste of time and money if the organizations being studied are not actually "world class." For example, Baxter Healthcare, a large manufacturer of medical and related equipment and supplies, decided to benchmark the human resource practices of some of the best organizations in the United States. Background research indicated two Baldrige Award-winning companies had world-class functions. After spending thousands of dollars in travel expenses and management time, the Baxter managers learned that, in many areas, these companies' HR functions were not as sophisticated as their own.

Many companies are considered to be prime candidates to benchmark, because they have good public relations functions—they know how to generate positive press about their organizations. Similarly, to make it easier for companies to identify who's best at a particular function, the American Productivity and Quality Center in Houston built a database that could be accessed before a benchmarking trip. The database was intended to provide detailed information on which organizations were world class in hundreds of areas. However, since companies nominated themselves for inclusion, in reality it may be a list of companies who consider themselves to be world class.

How to make benchmarking more effective. Keep the focus of benchmarking narrow—for example, studying how a company processes supplier invoices, or trains maintenance technicians. To increase the effectiveness of benchmarking, determine that an organization really is world class. Keep the focus on a specific process and do enough planning to establish the purpose of the trip, as well as the objectives to be achieved.

HOW TO BALANCE ACTIVITIES AND RESULTS

Organizations that successfully implement total quality management find that TQM pays for itself through revenue gains resulting from team projects and increased sales. These companies begin to see the returns within the first year. In the following discussion we'll address some of the actions that organizations can take to help ensure that the strategies and activities they implement will generate the revenue to fund efforts in the next stage of implementation.

Focus on Results

When implementing total quality management, organizations should focus on strategies and activities that result in improvements in all of the five major performance measures:

- Customer satisfaction measures.
- Operational measures.

- Quality measures.
- Financial measures.
- Employee satisfaction measures.

Any activity associated with TQM should be evaluated on its ability to positively impact result measures for the organizations. Activities that don't produce results in these areas should be modified or eliminated.

Focus on Process Control

Most quality problems are caused by variability in processes. Companies that can control processes and minimize variability can consistently produce products and services of high quality. ALCOA, the manufacturer of the rolled aluminum sheet used to make such products as beverage cans, knows a great deal about the process variables that need to be monitored and controlled to produce consistently high-quality rolled aluminum. Several hundred of ALCOA's scientists and researchers spent years isolating key variables that impact key product quality parameters. They identified the tolerances or upper and lower control limits for each key process variable. Control techniques then were designed and put into place to keep the process performance within the acceptable band of variability. ALCOA's approach to process control is a model for many of its customers. For example, wheelchair manufacturer Quickie Designs, Inc., has studied ALCOA's approach to quality and applied the approach in its own plant.

In manufacturing plants where processes are controlled by machines and computers, process control is particularly successful. However, in organizations where most of the work is done by highly skilled craftspeople, human behavior is the major variable. Since human beings cannot be programmed to respond the same way every time, the processes in which people are involved cannot be controlled with the precision that is possible in an environment like ALCOA's.

One company that does an excellent job of controlling processes that depend on human behavior is Ritz Carlton Hotels. The company defined more than 750 major processes, now collects data on each of them, and has control strategies in place to ensure that standards are met.

Checklists also are commonly used control strategies for behavior-based processes. For example, airline pilots are required to complete a preflight checklist before every flight. These checklists help to make certain that correct procedures are followed every time.

Focus on Indicators That Measure Performance Improvements

As pointed out in earlier discussion, many organizations mistakenly measure the amount of activity taking place, rather than whether the activity is yielding the desired result. In addition to tracking such factors as the number of hours employees spend in training and the number of people involved on teams, actual results should be measured, such as the effectiveness of the training and the improvements teams have made.

New England Telephone in Boston, one of the pioneers in developing effectiveness measures in training, does an excellent job of measuring the impact of training on job performance. Along with traditional measures of training hours and dollars spent, New England Telephone employs a variety of other techniques to evaluate training. Data include surveys of graduates and their supervisors, observations of job behavior, and measures of actual job performance. For example, the effectiveness of the training that installers receive is evaluated by measuring such variables as productivity and accuracy of installers before and after training.

Focus on Identifying Customer Needs

In several of the examples discussed earlier in this chapter, teams failed to identify customers' requirements. This oversight caused thousands of wasted hours and tens of thousands of wasted dollars. Before embarking on any process improvement activity, clearly identify the requirements and priorities of internal or external customers. This sounds so logical that it's difficult to imagine an organization's failure to do it. Yet, literally hundreds of process improvement activities are initiated without first performing this crucial step.

Focus on Long-Term Strategies

Organizations that consistently achieve anticipated bottom-line results focus on long-term strategies. While short-term measures often result in temporary financial improvements, concentrating exclusively on quarterly financial results and cost cutting will not ensure long-term profitability.

CONCLUSION

Many organizations do not experience an adequate return on their TQM investment. Usually, it's because they put too much focus on implementation activities, rather than on the results the activities should generate. To achieve the financial gains possible in the early years of the TQM effort, organizations should focus on strategies and activities that result in improvements in areas of customer satisfaction, operations, quality, and financial and employee satisfaction. All activities should be evaluated on their ability to impact these areas.

In this chapter, we pointed out many of the strategies and activities that lead companies to waste time and money, and we also discussed several actions that organizations can take to ensure their TQM activities positively impact result measures.

Tips, Tools, and Techniques
MEASURING TQM ACTIVITIES AND IMPROVEMENTS

Typical Measures of TQM Activities	*Suggested Improvements*
Number of hours of training per employee.	Percent of individual training/development plan objectives met.
Dollars spent on education/training.	Return on investment for training dollars.
Number of suggestions/ideas for improvement received.	Number of suggestions implemented or percent of suggestions implemented.
Number of teams (problem-solving, cross-functional, and so on) or percent of employees on teams.	Performance improvements derived from team projects.
Number of quality related awards/recognitions given out.	How employees feel about the awards and recognition items distributed and the fairness with which they are given out.
Number of speeches executives make on quality related topics.	Employee survey data on how they view executive commitment to TQM.
Number of layers of management eliminated to promote empowerment.	Employee survey data on the extent to which they have more authority/autonomy than in the past.

SUGGESTED READING

Brown, Mark Graham. "How to Guarantee Poor Quality Service." *Journal for Quality and Participation,* December 1990, pp. 6–11.

Crosby, Philip B. "Viewpoint." *The Quality Observer,* May 1992, p. 2.

Gilbert, Thomas F. *Human Competence.* New York: McGraw Hill, 1979.

Lee, Chris. "Do the Job, Hold the Service." *Training Magazine,* November 1989, p. 8.

Oren, Harari. "Ten Reasons Why TQM Doesn't Work." *Management Review,* January 1993, pp. 33–38.

Schaffer, Robert H., and Harvey A. Thomson. "Successful Change Programs Begin with Results." *Harvard Business Review,* January–February 1992, pp. 80–89.

PART II

WHY ORGANIZATIONS FAIL DURING ALIGNMENT

Organizations that are successful in Phase 1 are always eager to spread TQM throughout the organization. However, diffusing the success of a pilot into the organization brings a host of new obstacles. First is the issue of scale. Managing TQM in the entire organization requires different strategies and infrastructures than does a small pilot effort. Organizations often create "parallel hierarchies," with quality committees and teams that only add to bureaucracy. In their fervor, executives often imbue TQM with all the trappings of other "programs" that have come and gone, contributing to employee cynicism.

As TQM practices spread, the organization is confronted with systems and practices that conflict with TQM. Organizations quickly find that they do not have appropriate measurement systems in place and that some measures they do have promote poor quality practices. As organizations encourage teamwork, they quickly discover that their human resource systems focus on individual work, not teamwork.

During this phase, organizations must establish a credible, manageable implementation strategy and align several organizational systems that present immediate obstacles to progress. Organizations usually fail during Phase 2 due to four reasons.

Divergent Strategies

Employees in many organizations think that quality is separate from work; and, in their implementation strategies, organizations do much to reinforce this perception. Quality is addressed in quality improvement teams (QITs), not by the work group as a whole. These QITs often report to a tier of committees or to quality councils that are separate from the chain of command. Chapter Five explains how to integrate quality into the organization so it becomes part of every job all the time.

Inappropriate Measures

A basic tenet of management is that you get what you measure, and organizations tend to measure what is easy to count, not what is important. For example, many customer service representatives are judged on how many calls they can handle in an hour, not whether they satisfied customer needs. These measures are not balanced because many executives pore over financial reports and never ask for customer or quality data. And the measures are used to punish, not to learn. Chapter Six explains how to select appropriate measures to manage the quality effort.

Outdated Appraisal Methods

Traditional performance appraisal systems are damaging to teams and to quality. They encourage competition instead of cooperation, they reinforce traditional management practices instead of empowerment, and they focus on the manager's desires, not the customer's. Chapter Seven explains how to redesign performance appraisals to support quality and teamwork.

Inappropriate Rewards

Like traditional appraisals, most compensation systems focus on individual performance, and thus promote competition instead of teamwork. To combat this, organizations have implemented a variety of alternatives, including gainsharing and skill-based pay. However, these often only introduce new problems. In addition, executive compensation is rarely linked to quality. Chapter Eight explains how to design a compensation system that will support TQM.

Since Phase 2 spreads TQM into all the corners of the organization, success is predicated on aligning the connective tissue of the organization. In addition to establishing an effective diffusion strategy, organizations must address measures, appraisals, and rewards.

Chapter Five

Implementation Strategy

How to Avoid Making TQM a "Program" and Adding Bureaucracy

As many organizations enter Phase 2 of their total quality management initiative, they are faced with a common situation. Momentum generated by the initial success of the implementation has started to slow. Early results achieved by teams have deteriorated. Performance on most measures is essentially flat. Morale is down, and most employees perceive TQM to be just another program to deal with in addition to their regular work. Often this perception is reinforced by the organization itself.

In this chapter we'll address two ways that organizations reinforce the perception that TQM is a program, as well as discuss two additional practices that hinder its integration into the organization. We'll conclude the chapter with advice on how to integrate TQM into your organization and give five suggestions you can use to ensure it does not become a meaningless program.

WHY TQM LOOKS LIKE A PROGRAM

In organizations where the total quality management initiative is perceived to be a program, rather than a cultural change, TQM has not been integrated into the organizational structure. The quality initiative has a parallel, and separate, framework. The divergent structures are especially apparent in organizations where quality is addressed by quality improvement teams instead of by the work group as a whole, where these quality teams report to a tier of committees or quality councils isolated from the organization's chain of command, and where the initiative is under the direction of a separate group of individuals. The following discussion addresses two of the ways that organizations reinforce the perception that TQM is a program, and it examines practices that impede integration into the organization culture.

TWO WAYS ORGANIZATIONS REINFORCE TQM AS A PROGRAM

Seasoned American employees have seen many programs come and go. They are familiar with the distinguishing characteristics of a three-initial name and the accompanying slogans, signs, banners, bumper stickers, pins, and promises. Many employees persist in their belief that total quality management is just another program because the implementation has all the earmarks of a program.

Naming the TQM Implementation

Organizations perform considerable verbal gymnastics to ensure that employees do not mistake the total quality management effort for a program. Most of the rhetoric is designed to convince skeptical employees that TQM is not another program the company will discard in a year or two when something better comes along.

To strengthen the concept of TQM as a long-term change strategy, many organizations give the initiative a name, such as Vision 2000 or Continuous Improvement. Others go through extensive mental exercises to ensure that the name accurately reflects the intent of the initiative. One company calls its initiative "Market-Driven Quality" to reflect its change in emphasis from technology driven factors to customer driven needs. Another calls its initiative "Total Quality Leadership," believing nonmanagement employees would not support an effort with the word *management* in the title. Similarly, another company, which initially called its initiative "Total Quality Management," dropped the "management" in preference to simply "Total Quality." Yet another thought the word *quality* was too narrow to apply to its effort, which was aimed at improving productivity, cycle time, financial performance, and other types of results. Its program is just called "Total Management."

Naming the TQM initiative, however, is like the proverbial double-edged sword. On one side it gives the program an identity and gives meaning to the ensuing events. On the other side, giving the TQM initiative a name reinforces the perception that it is just another program. Many companies agree that a slogan or name, especially one that can be abbreviated to three initials, encourages employees to think that TQM is no different than any other program the company has started.

A label also allows employees to differentiate TQM, MDQ, or CI activities from activities they perform as part of their "real jobs."

Too Much Hype and Hoopla

Organizations that accompany the total quality management initiative with slogans, banners, buttons, and pins also reinforce the perception that TQM is a program. Employees remember that programs of the past were introduced with similar fanfare, and consequently, they see TQM as yet another program with slogans and banners.

The hype and hoopla foster distrust among employees and customers as well. The fanfare creates expectations, yet the slogans, signs, banners, and pins do nothing, by themselves, to improve quality or customer satisfaction. Unless many other actions are taken to bring product and service quality up to the level of those expectations, employees and customers will be disappointed.

Most people have been disappointed too often. For example, banners, cardboard signs, and tent cards—all proclaiming the commitment to quality and customer satisfaction—often are prominent in banks, stores, and restaurants where service is inferior. Slogans that proclaim "Quality Is Number 1" are common at companies whose products are mediocre. As a result, customers and employees have learned to distrust these slogans.

Employees similarly distrust the organization's slogans and declarations of values or intent. Many companies today have developed values that articulate the company's beliefs. These value statements usually include such words as *quality, meeting customer needs, safety, ethics,* and *happy employees.* Yet, from their perspective, employees do not believe these values drive the company's operation.

COMMON PRACTICES THAT HINDER THE INTEGRATION OF TQM

Many organizations maintain they don't have a total quality management program; rather, they have a culture change initiative. For example, General Electric insists it does not have a TQM initiative, and IBM calls its culture change initiative "Market-Driven Quality." Others insist that TQM is a macrochange strategy. The intent of these companies is to integrate TQM so completely into the organization that it is virtually indistinguishable. When organizations fail to meet that objective, it's often because of the manner in which they implemented the effort. Many designate a vice-president level position to oversee the initiative or a hierarchy of teams to implement it. Both of these practices have disadvantages.

Designating a Chief Quality Officer or VP of Quality

Ten years ago, the job of chief quality officer or vice president of total quality management was nonexistent. It's a relatively new position that many organizations, both large and small, have created for the person who leads the implementation of TQM. Typically, the person who fills the chief quality officer or vice president of TQM slot reports directly to the CEO, and he or she directs the total quality management activities of various business units in the company, develops and coordinates common approaches, and usually chairs the executive steering committee that over-sees the TQM effort.

Despite the fact that many of today's successful companies—including many Baldrige Award winners—have chief quality officers, the creation of a chief quality officer or vice president of quality is often a mistake. As mentioned in Chapter One, a major reason is that it shifts the responsibility for quality away from the organization's executives and relieves them of accountability.

The presence of a chief quality officer or vice president of quality also can lead to the creation of another staff department that drains overhead dollars. Although it often starts small, within a relatively short time the department may consist of the quality leader, a staff of several professionals, numerous part-time facilitators, and quality directors in each of the company's major business units. Additional meetings take place and extra paperwork and reports are generated. In essence, another staff empire is built and sustained.

Organizations that do not carefully select the individual for the chief quality officer position also create another problem—the individual chosen for the position is often the wrong person for the job. For example, many small and medium-sized companies choose an individual who held a middle management or technical professional position in the TQM area of a large corporation. Typically, this person knows little about the company or its business. He or she may be able to generate TQM activities but often fails to produce measurable results in bottom-line measures of quality and customer satisfaction.

Large corporations, on the other hand, usually promote from within to fill the chief quality officer slot. Again, however, the individual chosen is sometimes the wrong person for the job. Although there are notable exceptions, few companies select one of their superstars to fill the role of TQM vice president. More often the individual has held several staff management jobs, is loyal, and is well liked. The biggest challenge this person faces is that he or she has little or no clout with corporate and business unit executives and finds it difficult to achieve his or her objectives.

Creating a Dual Organization Structure

Numerous corporations have delegated the responsibility for managing the TQM implementation to a hierarchy of committees and teams. The structure (see Figure 5–1) usually includes a corporate steering committee of executives chaired by the chief quality officer. In many organizations, lower-level managers and total quality directors from the company's business units are on the steering committee. In more progressive companies, the CEO and all his or her direct reports are also on the quality steering committee. Regardless of membership, the role of the steering committee is to oversee the quality effort in the company, as well as to commit the company's resources and change policies or procedures as required to facilitate the implementation of TQM.

Beneath the corporate steering committee, there are usually steering committees at the business unit or division level. These committees, which are typically led by the head quality executive in the business unit, perform a similar function to that of the corporate steering committee. Organizations that use the Baldrige Award criteria as a guideline for their TQM effort often form another layer of committees—one for each of the seven categories in the Baldrige criteria. These committees often report to the corporate steering committee and include representatives from each of the organization's major business units.

Many organizations also create cross-functional process improvement committees or teams. These include individuals from several functions whose goal is to improve a major process that cuts across functions. For example, the new product design and introduction process committee

FIGURE 5–1
Typical Committee Structure

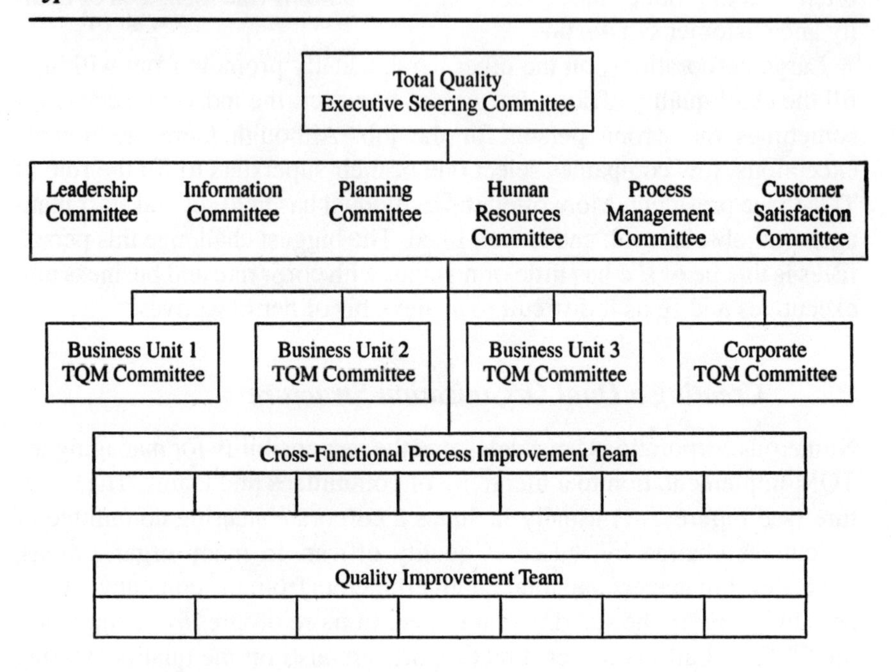

might consist of individuals from R&D, engineering, marketing, finance, safety, maintenance, and other areas.

Problem-solving or quality improvement teams are beneath the cross-functional teams. Many large companies have thousands of teams, consisting of employees from all levels working together to solve specific problems or to improve work processes in their own functions.

Although this approach to the TQM implementation is common and is used in numerous organizations, the disadvantage is that it creates a dual organization structure—a hierarchy of committees and teams parallel to the real organization structure, rather than integrated into it. This dual organization structure is one of the major reasons why TQM fails in most organizations: it impedes the integration of TQM into the day-to-day operation of the company. The dual structure also promotes the perception that TQM is a separate set of activities performed by a series of committees and quality executives—a program—rather than an attitude that should be integrated into every employee's job.

The committee approach to the implementation of TQM creates more committees, more meetings, and, in general, more bureaucracy in companies that already are riddled with it. In addition to the traditional organization structure, with its several layers of management, there is a TQM organization with a separate department and a hierarchy of committees and teams.

HOW TO INTEGRATE TQM INTO THE ORGANIZATION

The key to the success of any change effort is integration. The failure to integrate TQM into the organization structure is one of the main reasons the initiative fails in most companies. When TQM is seamlessly integrated into the way an organization operates on a daily basis, quality becomes not a separate activity for committees and teams but the way every employee performs his or her job responsibilities.

How to make sure total quality is not a program. Five suggestions to help prevent your total quality management journey from being another short-lived management program are listed below.

1. Don't give it a name or slogan, particularly one that can be abbreviated with three initials. If you need to call the effort something, give it a name that is consistent with your long-term business goals, such as "Vision 2000."
2. Eliminate the banners, slogans, buttons, posters, pep rallies, and other hoopla. Instead, concentrate on the essence of total quality management—process improvement.
3. Do one of two things: either don't appoint a chief quality officer or make it clear this is a temporary position that will be phased out after a few years.
4. Don't build a separate TQM department or organization that is responsible for the implementation of TQM. Every manager and employee should be measured on their quality and on how well they satisfy their internal and external customers.
5. Don't have separate committees to deal with quality related issues. Avoid forming a TQM executive steering committee. Rather,

make quality, customer satisfaction, and the implementation of TQM regular agenda items in executive and other management meetings that already occur.

Often this degree of integration is most easily achieved in a relatively young company or in a start-up operation. AT&T Universal Cards Services won the Baldrige Award in 1992, even though it was only a three-year-old company. One of the major reasons was that the Baldrige criteria was integral to how the organization was run. In fact, the former CEO, Paul Kahn, used the Baldrige criteria as a set of blueprints to design the organization—to define measures of performance, to develop strategic plans for the company, and to design processes that continually exceed customer requirements. Unlike many companies, AT&T UCS has never had a quality program. Employees are unable to make the distinction between "doing quality" and doing their "real jobs."

The same is true of Federal Express, a 20-year-old company that has integrated TQM so completely it is invisible to employees. Federal Express has never had a quality initiative or program, and yet the program has always ranked among the best in its industry for customer satisfaction.

Obviously, it is much easier to integrate TQM in a start-up company like AT&T UCS or in a company like Federal Express that has never had a quality program. It is difficult to change an old company that has always been strictly focused on the bottom line and the short term. It is especially difficult if the company has not had a near-death experience. But it is not impossible—and it is not necessary to make TQM another program to change the culture.

Changing the culture of a company simply means changing people's behavior. If employees' values and attitudes are changed, but those changes do not translate into changed behavior, the culture hasn't really changed. Changing behavior, as discussed in Part I of this book, requires a significant amount of training. It also requires follow-up measures to ensure that the concepts and skills learned in the training translate into changed behavior on the job.

Also necessary is to change many of the major systems that drive employee performance. The way employees are measured, what their priorities are, how planning is done, and how employees are compensated all play a role in changing behavior. There is more discussion on changing the culture of a company later in the book when Phase 3 is explained.

CONCLUSION

Many organizations on the threshold of Phase 2 of the total quality management implementation find that the effort has stalled. Usually, it's because the effort has not yet been integrated into the organization culture and into the management structure. In this chapter we examined the reasons why: TQM is perceived as a program, a perception many companies unwittingly reinforce; or TQM is isolated from the organization by the establishment of a hierarchy of committees. We conclude the chapter with some ideas on how to integrate TQM, along with five suggestions to help ensure it does not become a program.

Tips, Tools, and Techniques
LEARNING FROM OTHERS: A CASE STUDY—QUICKIE DESIGNS, INC.

Quickie Designs, Inc., of Fresno, California, is one organization that has successfully integrated the principles of total quality management into its organizational structure. Quickie Designs was started by a customer—Marilyn Hamilton, a charismatic entrepreneur and an accomplished amateur athlete. In 1979, Hamilton lost the use of her legs in a hang gliding accident.

The only wheelchairs on the market were slow, heavy, and ugly, and Hamilton had no intention of giving up tennis or the other athletic activities she enjoyed. She recruited two engineer friends, Don Helman and Jim Okamoto, who applied the technology used to build hang gliders to build a lightweight aluminum wheelchair with special wheels. The chair, which was painted neon pink, weighed about half as much as a normal wheelchair and was much faster. It looked more like a racing bike than a conventional wheelchair, and it folded to fit easily in the back seat of Hamilton's sports car.

Everyone who saw the chair asked about it. Soon, Hamilton, Helman, and Okamoto realized there could be a market for this type of wheelchair. In a shed behind Helman's house, they formed Quickie Designs and began building these fast lightweight wheelchairs.

In less than 13 years, Quickie Designs captured over 50 percent of the U.S. lightweight wheelchair market, and it now occupies a 150,000 square foot factory in Fresno. The product offering includes several lines of wheelchairs, as well as the original Quickie and a motorized version. Each chair is custom made for the user, based on his or her measurements, disability, and aesthetic preferences.

At the start, Quickie Designs identified the customer as the most important aspect of the business. As a customer herself, Hamilton kept everyone attuned to the needs of the customer. The company designs its products around the generic and specific needs of its customers. Traditional wheelchair manufacturers use the Model T Ford approach to product design: all wheelchairs are exactly the same and available with one finish—chrome. A key to Quickie's success is the ability to stay in constant touch with customers and use customer requirements to drive the product design process.

Quickie offers customers an infinite variety of features, colors, and dimensions for each chair. Just like people buying a car, Quickie customers can consider a number of basic models, select optional features, and choose the exact color they want. Quickie has designed a manufacturing plant that allows it to produce each chair to meet exact customer specifications.

Currently, 11 percent of Quickie's workforce is disabled, including six of the company's salespeople who are in wheelchairs. Since wheelchairs are sold through suppliers and therapists, this secondary group of customers is a critical part of Quickie's formula success.

In 1986, Quickie Designs was purchased by a much larger public company, Sunrise Medical Inc. of Torrance, California. Sunrise brought in Tom O'Donnell, a seasoned executive with 17 years' experience at IBM, as president. Although purchase by a big conglomerate often means the death of a small entrepreneurial company, this didn't happen to Quickie. Rather than make Quickie more like Sunrise, O'Donnell adapted to the Quickie culture, helping to further improve the organization without destroying its entrepreneurial spirit. O'Donnell also has shared the Quickie formula for success with other divisions of Sunrise.

Since O'Donnell came on board in 1986, Quickie has increased sales 25 to 30 percent per year. Hamilton is senior vice president of marketing and leads the company's efforts in support of disabled athletes such as wheelchair tennis champion Randy Snow. If you are ever in Fresno, visit the company. The people and products are truly exceptional. Quickie Designs is certainly a good model against which any company can benchmark itself.

SUGGESTED READING

AT&T Universal Card Services. "A Summary of the AT&T Universal Card Services Malcolm Baldrige National Quality Award Application." Jacksonville, FL: AT&T Universal Card Services, 1993.

Pickelle, Curtis. "The Quest—A Personal Story of How Three Individuals Conceived and Created the Lightweight Chair, Quickie, That Changed the Face of the Wheelchair Industry." *Q Magazine.* Los Angeles: Allied Health Care Publications 3–6, 1990.

Quality and Productivity Management Association. "Communicating Empowerment at Federal Express." *Commitment Plus* 8, no. 5 (March 1993).

Chapter Six

Measurement

How to Select Meaningful Quality Measures

Despite the unpredictability of human behavior, as a general rule organizations get what they measure from their employees. Nothing does more to keep an organization on a track of success than measuring the right variables and providing employees regular feedback on those variables. Conversely, measuring the wrong things can lead a company into trouble very quickly. In Chapter Four, we discussed how the quality movement often causes companies to measure activities, such as training and teams, rather than measure important results. In this chapter, we will talk about how organizations usually measure results, the types of variables they measure, and what can be done to improve measurement systems in most companies.

COMMON MISTAKES ORGANIZATIONS MAKE WITH MEASUREMENTS

The most common mistake organizations make is measuring too many variables. The next most common mistake is measuring too few. In the following discussion we'll address why organizations have a tendency to measure too much or too little, and we'll discuss several other circumstances in which organizations measure and use data in inappropriate ways.

Compiling Too Much Data

The databases that exist in most organizations were not built according to a plan or strategy—they simply evolved. Typically, the longer an organization has been in business, and the larger it becomes, the more data it collects and reports. One telecommunications company has 106 databases, each containing data on 75 to 100 separate performance indexes. This means the company can generate about 10,000 graphs. However, much of the data is unnecessary. No business is so complicated that 10,000 variables must be tracked, analyzed, and reviewed to determine how well the company is performing.

In most organizations, databases were not built as the result of a thoughtful determination of exactly which variables need to be tracked. Databases usually have evolved indicator by indicator, over a long time. As the evolution progresses, most of them become cumbersome, producing increasing numbers of charts and graphs that managers and employees either ignore or use ineffectively.

Failing to Base Decisions on Data

One of the premises of the quality movement is that product, process, or service can't be improved unless it can be measured. As a result, there is no shortage of data in most organizations. There is, in fact, so much data being generated that many executives and managers are overwhelmed with information and find it difficult to make decisions. Often, managers don't take time to review several pages of charts. Those who do often find the data too complicated to analyze. When frustrated by too much data—or too complex an array—many managers opt to make the decision based on intuition or past experience. This, of course, defeats the purpose of collecting the data.

According to a group of Japanese workers involved in a project to teach American workers in a Toyota plant to solve problems, the major difference in the American and Japanese approaches to problem solving is the way data is used. They said:

> Both Japanese and American companies collect a great deal of data. The difference is that we trust the data and use it to make key decisions. Americans don't trust the data unless it goes along with what their gut tells them. They make decisions based on intuition and past experience, not data.

When data and intuition conflict, managers tend to favor intuition. They often justify the decision by explaining how the data collection or reporting was flawed. What they usually mean is that the data are too hard to interpret, or there is too much information to assimilate.

Unspoken Measurements

Every company has an unspoken code of behavior that an individual is measured against. In some companies, the code is "keep quiet, do your job, don't challenge the boss, and do what you are told." In other companies, the code is "be aggressive, challenge everyone, take risks, and implement new practices." An individual's ability to learn the organization's code of behavior is still one of the variables that determine how far up the ladder he or she moves.

Despite value statements that declare what the organization professes to believe, in many companies the unspoken priorities are to achieve financial targets, avoid making waves, and work long hours—at least 10 hours a day during the week and another half-day two or three Saturdays a month. Factors like the following are given preference:

- How an individual dresses.
- How many hours an individual works.
- How an individual behaves in meetings.
- How well an individual gets along with superiors.

Obviously, no organization collects statistics on how managers dress or on what time they arrive in the morning. However, people do keep track. It is well known in certain circles who usually arrives late and who always challenges superiors in meetings.

Incomplete Measurements

Earlier we discussed the mistake of measuring too many variables, which is particularly common in large corporations. Some smaller companies tend to err on the other side of the scale. They measure too few key variables to provide a complete picture of the health of the organization.

Often, smaller organizations measure only a few of the key financial variables to monitor how the company is doing. If the numbers are good, it is assumed that other factors are in line. However, short-term financial results are not as strong an indicator of success as are longer-term measures, such as customer satisfaction, employee satisfaction, and market share.

Many smaller companies, which concentrate on the key variables they believe are most crucial and ignore longer-term indicators of success, do so because they don't have the time or resources to measure the broad range of indexes that are measured in larger companies. However, having too little data on which to base decisions is as problematic as having too much data.

Inconsistent Measurements

In many organizations, much of the data collected is conflicting, inconsistent, and unnecessary. To be meaningful, all the data collected in an organization should interrelate and lead to the overall measures of the health of the whole organization. For example, employee measures of performance should feed into the manager's measures of performance, which should feed into the department's measures of performance, which feed to the division's measures of performance, which lead to the organization's measures of performance. In other words, all data should lead to some ultimate measure of success for the company.

Inconsistencies in data and unnecessary information can be identified with the use of "Organizational Mapping," a technique pioneered by consultant Geary Rummler. The approach, which involves drawing a logic trail for the data collected to determine where it leads, does work, but it is often a long and difficult process. Many organizations prefer to rewrite faulty databases from scratch, starting with a good model of what should be measured at the top, and working downward.

HOW TO SELECT APPROPRIATE MEASURES

Every organization, regardless of the type of business it is engaged in or its size, needs to measure the same types of variables. These variables are:

- Customer satisfaction measures.
- Financial measures.
- Product/service quality measures.
- Employee satisfaction measures.
- Operational measures.
- Public responsibility measures.

Every individual measurement index in an organization should fall into one of these six categories, and all organizations with more than a handful of employees need to have data on each category of measures. The weight or priority given to each of these major types of measures may differ considerably from organization to organization, however. For example, during a cash flow crunch, an organization's priority may be to generate sales. Another organization may be performing adequately in areas of satisfying customers and generating revenue, but it may have priorities in such areas as dealing with environmental and other public responsibility issues. In the following discussion we'll look at these six areas in more depth.

Measuring Customer Satisfaction

It is difficult to find an organization that has not started to collect some type of data on customer satisfaction. Companies that measured customer satisfaction by keeping track of complaints or of lost business now take a more proactive approach. Instead of leaving comment cards on tables in restaurants or on dressers in hotel rooms, many survey customers by mail, telephone, in person, or in focus groups. These companies have found that relying on customers to fill out comment cards does not provide a representative sample. Usually, only customers who are very angry or especially delighted take time to fill out comment cards. Although many companies still use the cards—they want to know if a customer is very angry or very happy—they realize that this data is not a reliable way to measure overall levels of customer satisfaction.

Collecting data on customer complaints. Many organizations supplement data from comment cards by tracking customer complaints. One paper company even uses customer complaints as the major measure of customer satisfaction. However, while it is important to track complaints, and to have a system to respond to them promptly, complaints are not an adequate measure of customer satisfaction.

One reason is that some consultants suggest only 1 in 25 unsatisfied customers actually complain. Of the remaining 24, most simply take their business elsewhere. One company counts every customer complaint as 25, and it graphs the data so employees can see the real picture, rather than the tip of the iceberg.

Another reason why it is unwise to use complaints as a measure of customer satisfaction is that most complaints are made informally and, as a result, are not recorded. Generally, the only complaints companies record are those submitted in the form of a letter or phone call to the customer service department. It's doubtful comments made to such personnel as hotel desk clerks, restaurant servers, and sales representatives are recorded in the organization's databases.

Types of customer satisfaction surveys to avoid. The standard customer satisfaction survey asks respondents to rate the quality of the company's products or services on a point scale similar to the one below.

5	*4*	*3*	2	*1*
Outstanding	Very Good	Satisfactory	Fair	Poor

The ratings are tabulated and averaged to yield a summary statistic that may be, for example, a 3.8 on a 5.0-point scale. A large real estate company measures customer satisfaction on a scale much like this. Trade magazine ads claim that the company has 96 percent customer satisfaction. However, when calculating customer satisfaction, the company considers 2 to 5 on the scale as satisfied. The company does not track the percentage of 2s, 3s, 4s, and 5s. This approach to measuring and calculating customer satisfaction leads to mediocrity, not to delighting customers. It is possible to get lackluster ratings—all 2s and 3s—and still end up with a score of high customer satisfaction.

Another way to stack the deck to ensure favorable ratings is to ask questions that elicit a positive response. For example, when customers renew driver's licenses and automobile plates, a department of motor vehicles asks if the amount of time they waited in line was reasonable compared to their past experience at DVM. Most people have adapted expectations to past experience. If the wait is less than they've experienced in the past, they probably will respond positively to the question. Using the phrase "reasonable, based on past experience" but allowing only a yes/no answer further biases the survey toward a positive response. However, in its data, this department of motor vehicles reports about 65 percent of its customers rated the wait at DMV as reasonable.

Designing a customer satisfaction survey to promote excellence. Another company measures customer satisfaction with a system similar to the one previously described, but its system promotes excellence. Solectron, a circuit board manufacturer, and a 1991 Baldrige Award winner, has customers give it a letter grade, just like the traditional grade students receive in school. In the standard calculation, an A is worth 4.0, a B 3.0, a C 2.0, and so forth. On Solectron's five-point scale, which is designed to promote A-level performance, GPAs are calculated in a non-traditional manner. According to its CEO, Dr. Winston Chen, "Average work is worth nothing in our company. Therefore, a C rating from a customer, meaning satisfactory, is worth zero on our scale."

Solectron considers a grade of D as a negative 150 points, and an F a negative 300 points. A grade of B is worth 50 points and an A 150 points. This scoring/grading scale encourages A-level performance and gives no credit for customers who rate quality as "satisfactory."

Generic customer survey questions. Although customer surveys should be tailored to each organization, some questions probably should be included on all questionnaires; these are:

Rating Questions

1. Overall, how would you rate the quality of our products/services?
2. How would you evaluate our responsiveness to your concerns/problems?
3. How easy are we to do business with?

Open-Ended Questions

4. What products/services should we offer that are not currently offered?
5. What is one thing we could do to improve your level of satisfaction?

Hard measures of customer satisfaction. Although it is necessary to obtain customer input about products or services, such mechanisms as surveys, focus groups, and customer interviews provide only about half the information an organization needs to determine levels of satisfaction. Customers' behavior, which is far more revealing, provides the remainder. For example, a customer may give his or her car a mediocre rating on a survey, yet trade it in on a newer version of the same model. This customer's buying behavior indicated enough satisfaction with the car to buy another one.

Consequently, a good customer satisfaction measurement system includes soft measures from such mechanisms as surveys, focus groups, and complaints, combined with hard measures that are based on behavior.

FIGURE 6–1
Customer Satisfaction Index (percent)

• Telephone survey ratings	25%
• Focus group ratings	5
• Complaints	15
• Repeat business	40
• Market share	15

Typical hard measures include gains and losses of customers, repeat business, increases or decreases in volume of business, or market share. Some leading companies calculate a customer satisfaction index, or CSI, that includes a mix of hard and soft measures. Each individual index is given a weight, based on its importance, and the degree to which it is a good indicator of success. An example is shown in Figure 6–1. Note that 45 percent of the points in the CSI are based on soft measures and 55 percent are based on hard measures. While not a requirement, a 50/50 or a 40/60 mix is suggested.

The key to any good customer satisfaction measurement system is to use a variety of different measurement instruments and methodologies to get a well-rounded view of how customers feel about products or services. Some corporations with 20,000 employees measure customer satisfaction by surveying their top 100 customers by mail once a year. Successful companies, on the other hand, put a great deal of time and money into measuring customer satisfaction. Stew Leonard's Dairy, a company of about 400 employees and with two stores, uses eight different methods to measure customer satisfaction. Among them are phone surveys, mail surveys, focus groups, parking lot interviews, and in-store interviews by management and by checkers. This company puts far more resources into measuring customer satisfaction than do many companies 100 times its size.

Measuring Financial Performance

Most companies already do a good job of measuring financial performance. For the purposes of this discussion, we will assume that most organizations already have adequate financial measures in place.

Every employee, from the CEO down, needs to be measured on some financial dimension of his or her performance. A number of organizations measure financial factors only at top-management levels, because these people have the most control over these variables. This is a mistake because, to some degree, every employee has the potential to impact the organization's finances in a positive or negative way.

Measure people on financial factors they can control. It is common for organizations to measure people on financial variables over which they have little or no control. Financial measures should be indexes that the person or team does not necessarily totally control but does have a fair degree of influence over. It is inappropriate, for example, to measure an individual engineer on how well the engineering department stays within budget. On the other hand, it is appropriate to measure a sales representative on the gross margin produced from a sale. A sales representative does not have total price flexibility—he or she must work within guidelines or limits—but the sales representative often does have some control over the price that he or she charges for the product or service.

Select only a few key financial measures. Financial variables are just one of the areas in which every employee should be measured. To prevent data overload, financial measures for each individual should be limited to two or three. For example, the two primary financial indexes for a sales representative might be gross sales revenue and gross margin. In many organizations middle and upper-level managers are evaluated on more than a dozen measures of financial performance each month. These same employees also are measured in areas of quality, operations, and customer satisfaction. Managers should select a few key financial indicators to watch and control, rather than attempt to control all of them. A word of caution, however. The disadvantage of this approach is that the manager may disregard the most important financial measures in favor of the ones on which he or she can perform well.

Find nontraditional ways to measure financial factors. Progressive organizations have gone beyond the standard financial measures of return on assets (ROA), profit, and sales and are using nontraditional approaches to track financial factors. Many companies use an approach called "Activity-Based Costing," or ABC, to track the cost of various activities. ABC is based on the premise that understanding the true cost of activities performed in the organization will result in reductions in activities that waste time and money.

Another innovative financial approach originated with Phil Crosby—the cost of quality, or COQ. Crosby's premise is that organizations spend significant amounts of money to detect and fix mistakes, which are the true costs of achieving good quality. Crosby suggests it is almost always less expensive to prevent mistakes than to find and correct them. Organizations that measure the cost of quality become aware of these costs and can work to control them.

Devotees of Crosby arduously measure the cost of quality throughout their companies, and in some organizations the practice is out of hand. What began as an exercise to demonstrate the importance of doing things right the first time has turned into a massive report of COQ figures. According to a quality manager for a major corporation, one of the best moves his company made this year was to discontinue the cost of quality measurements.

> We had a department of 14 full-time people compiling data and issuing reports on the cost of quality in our organization. Most managers ignored the reports because we have gotten to the point where we now have a prevention-based approach to quality. We eliminated the COQ measurement department and put those 14 people to work generating data that we really need to make business decisions.

Measuring the cost of quality in this company helped raise awareness of the cost of rework, which prompted employees and managers to redesign processes to prevent problems, rather than to detect and correct them. When a more preventive approach is implemented, companies often discover it is unnecessary to measure the cost of quality.

Measuring Product/Service Quality

All organizations, whether an entire company or an individual department within a company, produce products or services. Therefore, all organizations should have some measures of the quality of the products and services that they produce. Product/service quality measures are different from customer satisfaction measures. They are not based on how the customer behaves or what the customer says. Product/service quality measures are the internal data collected on the quality of products or services before they reach the customer. This data may be collected by the department that actually produces the products or delivers the service, or it may be collected by a separate quality inspection function/department. The important factor is that the measurement is done by company employees, not by customers.

Dimensions of quality. Quality often is considered to be simply a measure of accuracy, but it is much more than that. Although accuracy, or the absence of defects, is important with almost every product or service, sometimes other dimensions of quality are equally, or even more, critical. In determining which aspects of product/service quality to measure, one of the first steps is to define which quality dimensions of your products/services are most important to customers.

Quality can include measures of:

- Accuracy.
- Completeness.
- Conformance.
- Innovation/novelty.
- Class.

Frequently, *accuracy* is the important dimension of performance. For example, in employee drug testing, accuracy is the most critical performance dimension, since careers can be ruined by the results.

The second type of quality measure is *completeness.* This dimension relates to whether all the requirements or components are present in a product or service. For example, omitting a zip code on an invoice is a completeness problem, not an accuracy problem.

A third dimension of quality is *conformance.* Sometimes a product or service is accurate and complete, but the form does not follow established guidelines or standards. For example, it may be important that all documentation for product specifications be in a specified format so all specs are consistent. The information may be accurate and complete, but, if it is in the wrong format, it is a conformance problem.

Innovation or *novelty* is another dimension of quality that may be important for products or services. To illustrate, in the products produced by researchers, or by art directors in an advertising agency, innovation or novelty is one of the most important dimensions. In the manufacture of car tail pipes, innovation may be inappropriate. Innovation or novelty cannot be measured as objectively as, for example, the number of defects, but sometimes it is much more important.

The last dimension of product/service quality to be considered is a dimension frequently called "*class.*" It often relates to aesthetics or to what people in the real estate business call "curb appeal." If, for instance, the administrative services department putting together overhead transparencies for a speech turns the boring originals into exciting visuals, this is a measure of class. Class measures of quality are often very subjective but, like innovation, they are sometimes more important than objective measurable dimensions, such as accuracy.

Measures of Employee Satisfaction

Employee satisfaction measures often are missing from an organization's database. Some of the more forward-thinking companies like Marriott and Federal Express believe that employee satisfaction is the first step to customer satisfaction. Employees happy with their jobs, their compensation, and their organization do more to delight customers. Unhappy employees do just enough to avoid losing their jobs. This approach is called "the customer comes second." To delight customers with exceptional products and services, the employee is the first customer the organization has to satisfy.

Hard and soft measures of employee satisfaction. As with customer satisfaction, employee satisfaction measures should combine a variety of hard and soft indexes. Many large corporations limit the data they collect on employee satisfaction to an employee morale survey conducted every two years. However, this is not enough. The soft aspects of employee satisfaction measures should come from surveys, focus groups, and other mechanisms that obtain details on the way employees feel about the company, their jobs, and how they are treated.

Employee satisfaction surveys should be designed for each organization or conducted by a professional survey organization. Here are basic questions the survey should include:

- To what extent do the leaders of this company actually live by the values the company espouses?
- How committed are executives to the total quality management effort?

- Do you have ample opportunities to contribute to improving the company?
- Do you have more autonomy and decision-making authority than you did in the past?
- Have you received an adequate amount of training to do your job well?
- Is there an atmosphere of trust and open communication in the company?
- Are you proud to work for this company?

Many large companies also conduct exit interviews when an employee leaves the organization. This data seldom is summarized or recorded, despite the fact that exit interviews are an excellent source of feedback. Employees who are leaving the company are often more truthful than those responding to a morale survey.

Hard measures should be based on employees' behavior, not on their opinions or feelings. Some excellent hard measures of employee satisfaction are:

- Turnover.
- Requests for transfers.
- Grievances/complaints.
- Absenteeism.

The best way to measure employee satisfaction is to compute an employee satisfaction index, or ESI, similar to the customer satisfaction index described earlier in this chapter. About half the points should be based on soft data from morale surveys, focus groups, and exit interviews. The other half should be based on hard measures, such as those listed above.

In most organizations, compiling data on employee satisfaction is not enough. Convincing executives to pay as much attention to these data as they do to financial and customer satisfaction data is critical. Executives in most organizations review financial and operational data every week, and, typically, they review customer satisfaction data several times a year. However, unless there is a major problem such as a strike, they never look at employee satisfaction data; it is just not a priority for most executives.

One way to raise the priority of employee satisfaction data in executives' minds may be to demonstrate the correlation between employee satisfaction and more important measures, such as bottom-line profits. Another may be to tie this measure into executives' bonuses like Federal Express does. Although this gambit gets attention very quickly, most companies are reluctant to tie compensation to anything other than financial measures.

Operational Measures of Performance

Operational measures, like financial measures, usually are part of databases in most companies. This type of data is critical and is different from measures of quality or customer satisfaction. Operational measures are essentially those elements of an operation that don't fit into the other five categories. Some common operational measures are:

- Productivity.
- Cycle time.
- Scrap/waste.
- Energy efficiency.
- Proposal wins/losses.
- Raw material usage.

In many organizations, operational measures are called "process measures." These measures—if defined correctly and controlled—lead to the production of quality products and services. Often, however, the variables designated as operational measures are not the key process measures that need to be controlled to produce the quality products and services that customers want. For example, sales representatives might be measured on call frequency, even though there is no documentation to show that call frequency correlates to either sales volume or customer satisfaction.

Frequently, process or operational measures are based on superstition, tradition, or ease of measurement. Service companies often measure factors that are easy to track, but these may not be related to customer satisfaction. For example, a major hotel chain measures whether employees call guests by name when addressing them. Most hotels also measure whether the end of the bathroom tissue is folded into a precise arrow shape. Because it is unlikely that customers stay at particular hotels as a result of these measurements, focusing on the wrong details can waste a great deal of resources in a company.

Measures of Public Responsibility

Every year *Fortune* magazine conducts a survey to identify the most admired corporations in America. The ratings are based on the opinions of American executives, who are interviewed and asked to name the companies they admire most and least. Part of the assessment is based on financial results and good management strategy. Another part is based on the company's reputation for ethics, environmental protection, and corporate citizenship. These latter measures of public responsibility often are missing from the databases executives review on a regular basis. Like employee satisfaction, this type of data is only gathered when there is a crisis.

Measures of public responsibility differ from the previous five types of measures in that not every employee needs to be measured in this area. Public responsibility measures apply more to the entire organization than to the individual employee. Although a few organizations include these types of measures for all employees—for example, bank branch managers and other customer contact personnel have civic responsibilities as part of their overall job performance measures—they are not necessarily required for a sound database.

MEASUREMENT TECHNIQUES: COUNTING AND JUDGMENT

All measurement systems are based on two techniques of measuring performance—counting and judgment. Counting is the more objective of the two approaches. It is preferred over judgment, assuming that the things to be counted are the most important dimensions of performance. Many times they are not. For example, R&D functions often measure such considerations as the number of articles published by researchers or the number of patents received.

Counting. Counting as a measurement technique can be used with all six categories of measures discussed to this point. For customer satisfaction, the number of complaints received can be counted; defects or

FIGURE 6–2

Judgment	Objective	Subjective
Ranking	1 Specific Criteria	2 Opinions
Rating	3 Specific Criteria	4 Opinions

problems in products can be counted; the number of employees who resign can be counted as a measure of employee satisfaction; and the number of violation notifications received from OSHA and the EPA can be counted as a measure of public responsibility.

Other aspects of performance cannot be adequately measured by counting. These measures of performance require judgment. For example, when customers or employees are surveyed, they are asked to make a judgment about some aspect of performance.

There are four techniques for using judgment to measure performance, as shown in Figure 6–2:

1. Ranking specific criteria.
2. Ranking based on opinion.
3. Rating using specific criteria.
4. Rating based on opinion.

Ranking based on specific criteria. In a ranking system based on specific criteria, the evaluator ranks performance, people, or products from best to worst, based on one or more specific criteria. An example of such a system is ranking salespeople from best to worst in both sales volume and gross margin generated from sales.

This approach is most useful when there are no specific standards against which to judge performance. One person or performance must be compared to another person or performance to determine rank. In the Malcolm Baldrige National Quality Award, a ranking system is used to separate the finalists from the winners. Since just two companies can win in each of the three categories (service, manufacturing, and small business), finalists must be ranked from best to worst to determine the winners.

Ranking based on opinions. The second way of using judgment as a measurement technique is to rank performance, people, or products based on opinions. This type of system is a very effective way to determine which features of a product or service are most important to customers. For example, a manufacturer of cameras might list the following features, asking customers to rank them from most to least important:

Feature	*Rank*
Low price	___
Auto-focus	___
Warranty	___
Compact size	___
Reliability	___
Auto film advance	___

Of the four approaches to using judgment, ranking based on opinions is the approach most often used inappropriately. For example, one organization distributes awards to employees who suggest ways to improve company performance. Suggestions that are implemented are rank ordered from best to worst by a committee. There are no criteria for this ranking—each committee member selects what are, in his or her opinion, the 10 best ideas. The top 10 are then rank ordered from best to worst, and awards are given to those who submitted the ideas. The awards, which can be as high as $10,000, generate considerable interest. However, employees suggest that the system would be more equitable if criteria were identified to evaluate the best ideas.

Rating based on specific criteria. The best of the four techniques for using judgment as a measurement approach is rating performance based on specific criteria. This approach assumes there is an absolute standard of performance that is used for comparison. Rating based on specific criteria often is used by quality inspectors to assign ratings to different dimensions of product quality. These ratings are based on given scales that score various aspects of the product's characteristics or performance.

The first stage review in the Baldrige Award process also uses this system. Examiners score written award applications that are based on the response to each of the numerous areas addressed in the criteria.

Rating based on opinions. The last approach for using judgment as a measurement technique is to rate performance based on opinions, rather than based on specific criteria. One person or product is not compared to another. Rather, the comparison is to a predetermined standard.

This approach is slightly better than ranking based on opinion. The disadvantage is that the standards used for rating vary considerably from individual to individual. Many teachers and professors use this approach, grading each paper according to their own standard of what an "A" paper should be.

REDESIGNING THE MEASUREMENT SYSTEM

The databases and measurement approaches used in most organizations require alteration to promote the implementation of a more customer-focused culture. If the database already includes data in all six categories discussed in this chapter, and there are not too many variables being measured, this alteration could be simply a matter of adding measures and deleting others—a minor tune-up.

In most organizations, however, it will take more than a minor tune-up to update the measurement systems—it will probably take a major overhaul. This is especially true in larger, older companies, where the project may take several years and require a great deal of work. The following discussion covers many of the factors that need to be addressed when redesigning a measurement system.

Begin at the Top

The hierarchy of measures should begin at the top-management level and cascade down to lower levels. Designing the measurement system in this manner helps assure that all measures are interrelated, and that only the most important variables are tracked and included in the database.

Federal Express has a system in which every measurement index relates to one of three categories of variables: people, service, or profit. The people measure relates to employee satisfaction. Service is a measure of internal or external customer satisfaction, and the profit is obvious. We'll discuss the Federal Express system in more depth in Chapter Ten.

When a set of criteria to use in selecting and "de-selecting" indexes for the database has been identified, the next step is to list overall indexes for each of the six categories of data we've talked about in this chapter:

- Customer satisfaction.
- Financial performance.
- Product/service quality.
- Employee satisfaction.
- Operational measures.
- Public responsibility.

After brainstorming the macro-measurement indexes for each of these six categories for the whole company, the next step is to eliminate the indexes that don't meet the criteria, to combine separate indexes into ratios, and to otherwise shorten the list. For example, a statistic on overall operating cost might be combined with personnel costs to determine a ratio.

Another approach is to determine how several individual measurement indexes can be combined into a single score, such as described with the customer satisfaction index or with the employee satisfaction index. These summary statistics make data more manageable. The ideal database has no more than five individual statistics in each of the six categories of data. This means a total of about 30 individual measures is the maximum number of macro measures in the database. A database of 15 to 20 indexes is even better.

In most organizations, financial and operational indexes will not change significantly from those currently being measured. Most manufacturing companies also have very good product quality data, and changes in this area are unlikely as well. Typically, in most organizations, much of the effort will be expended on measures in the other three categories—the areas of employee satisfaction, customer satisfaction, and public responsibility measures.

When an appropriate list of macro measures has been identified for the entire organization, the next step is to create a matrix to identify the impact that each of the functions and business units in the organization has on each of the overall measures. This matrix helps determine which functions will have the primary and supporting responsibilities for each of the macro indexes. For example, the R&D function might have the primary responsibility for measures of new product success; however, other functions are involved as well. Marketing, manufacturing, and other functions have support or secondary responsibility. This measure might be given a weight of 25 percent for R&D personnel and only 5 percent for other functions involved in the design process.

When macro measures for the organization are complete, the same task is performed for each of the major functions or business units, or both. This process is continued in a cascading fashion, until a hierarchy of measures is established from the CEO level down to the individual contributor level. However, when cascading measures down through the organization, extreme care must be taken to ensure they are not pushed down too far.

Measures can be devisive, and it is often unnecessary and unwise to apply them to individuals or small groups. Measures at this level are only beneficial when the individual or small group completes a clearly defined output, independent of other groups.

For example, when the performance of shift operations are measured by shift, the shift workers quickly learn to postpone maintenance, cleanup, and difficult or disliked jobs until the next shift. Resentments soon swell, undermining all potential for shared learning and cooperation. It is more useful to focus measures and attention on the combined results of the larger work unit. Likewise, goals and rewards should focus on the larger work unit as well. At the shift or individual level, fine-detail measures should be used only for learning. If one shift develops a better process, the measurement system should encourage the team to share, not hoard, the development.

CONCLUSION

Many organizations collect either too much or too little data, and often they collect data that is wrong, inappropriate, or unnecessary. In this chapter, we discussed at length the six categories of performance in which data should be collected.

Some organizations already have an adequate measurement system that addresses all performance categories. Most organizations, on the other hand, need to reevaluate the indexes they are tracking and to completely overhaul their measurement mechanisms. This chapter concluded with advice and insight into both processes.

Tips, Tools, and Techniques
FINE-TUNING YOUR MEASUREMENT SYSTEM

Many measurement systems don't need a complete redesign. All that's required is a tune-up to ensure that measurements promote the implementation of a customer-focused culture. The following guidelines will help you to assess your existing database and decide what aspects need fine-tuning.

1. Make sure no employee or team is measured on more than 15 individual indexes of performance.
2. Make sure every employee or team has some measure of internal/external customer satisfaction.
3. Develop a customer satisfaction index (CSI) based on a mix of hard and soft measures of customer satisfaction.
4. Develop sound measures of employee satisfaction based on a mix of hard and soft data.
5. Assign goals, objectives, or standards to each performance index.
6. Establish goals, objectives, or standards based on customer requirements and benchmarks, and strive for continuous improvement.
7. Identify and eliminate the "disconnects" in your database—measures that don't lead to higher level measures of performance.
8. Assign weights to each of the categories of measures and the individual performance indexes based on the priorities for the organization and unit.
9. Question traditional activity and process measures to ensure that the variables measured are highly correlated with quality products/services.
10. Design a system to evaluate and continuously improve your database and data collection reporting processes.

SUGGESTED READING

Brown, Mark Graham. "You Get What You Measure: Engineering a Performance Measurement System." *Performance & Instruction Journal,* May/June 1990, pp. 11–16.

Garson, Barbara. *The Electronic Sweatshop: How Computers Are Transforming the Office of the Future into the Factory of the Past.* New York: Simon & Schuster, 1988.

Hayes, Bob E. *Measuring Customer Satisfaction: Development and Use of Questionnaires.* Milwaukee: Quality Press, 1992.

Hronec, Steven. *Vital Signs: Using Quality, Time, and Cost Performance Measurements to Chart Your Company's Future.* AMACOM, 1993.

Sellers, Patricia. "Companies That Serve You Best." *Fortune,* May 31, 1993, pp. 74–88.

Chapter Seven

Appraisals

How to Redesign Your Performance Appraisal System to Support Teams

As organizations implement TQM and empowerment, they quickly discover that their traditional performance appraisal system becomes a significant obstacle. Instead of focusing on the customer and promoting teamwork, these systems focus on the desires of the boss and promote competition. In this chapter, we will examine why traditional performance appraisals inhibit quality and teamwork. Then we provide a foundation for redesigning your system so it supports TQM values and objectives.

PERFORMANCE APPRAISALS AND GOAL SETTING

One of the principle purposes of traditional performance appraisals is to set performance goals against which an individual can be measured. Conventional wisdom maintains that, if you set goals, you improve performance. Imagine for a minute that you had four teams, each performing the same interdependent task. In one team you assigned individual goals, in another just a group goal, in the third both an individual and group goal. The fourth team was given no specific goal at all. Which team do you think would perform the worst?

Many people would answer the team with no goal. According to research that replicated this situation, they would be wrong. [1] In fact, the team with only *individual* goals consistently performed the worst. The team would have been better off with no goal at all! So much for conventional wisdom.

Interestingly, of the remaining three goal situations (i.e., no goal, group goal, and individual plus group goal), the teams with both group and individual goals generally performed the best; but the difference in performance was not statistically significant. This research implies that traditional appraisals (which emphasize individual goals) are counterproductive when employees are interdependent.

Why does this research contradict other studies? Most studies on the effect of goals use independent tasks where people do not have to cooperate. However, the days of the individual contributor—if they ever existed—are gone. In the complexity of today's business, our employees must collaborate. As a result, we need to understand the dynamics of goal setting, measures, and feedback within the context of teamwork, not individual work.

Most performance appraisal systems today are still based on an individual-contributor model. Managers have a series of one-on-one meetings with subordinates to establish annual goals, to give feedback on performance, and to mete out rewards. This system is like a conductor coaching each member of an orchestra privately and then expecting the performance to go well.

In exasperation, some organizations have ditched their entire performance appraisal system (to the consternation of their legal counsel, of course). This is equally inappropriate, but not so much for legal reasons. To improve, any system needs feedback. That's why the C is in Shewart's quality improvement PDCA cycle (plan–do–**check**–act). Regardless of its flaws, any performance appraisal process sanctions reflecting on the past and setting stretch goals for the future. To succeed in turbulent times, we need more and more frequent feedback, not less. We simply need to redesign the process to support a customer-focused, team-oriented, flexible workplace.

TRADITIONAL PERFORMANCE APPRAISALS DEFINED

How do you know if you have a traditional performance appraisal system? Here is a generic description. See how your system compares.

The process revolves around a formal performance appraisal meeting in which the employee and manager discuss the employee's past performance and set goals for the new year. Between formal appraisals, managers are supposed to keep documentation on employee performance and to periodically give informal feedback. Managers use their documentation to com-

EXHIBIT 7–1
Typical Performance Appraisal Form

Performance Appraisal for ____________________

Last Year's Goals	Rating
________	________
________	________
________	________
________	________
________	________

Other Factors:	Rating
Safety	________
Teamwork	________
Attendance	________
Dependability	________

Overall Rating:

Next Year's Goals	Weight
________	________
________	________
________	________

Development Plan

Disclaimer: Signing does not indicate agreement.

________	________
Manager	**Employee**

plete the required performance appraisal form. A typical form is shown in Exhibit 7–1.

The form typically includes space to write critical job duties or goals and performance standards. Managers are expected to assign a rating to the employees' performance as measured against the standards. Near the bottom of the form is the overall rating of the employee—usually on a five-point scale. This rating is so often tied to compensation that the amount of merit increase permitted is determined by this overall rating—a system commonly referred to as "pay for performance." Some forms have traits or general performance factors to rate, such as teamwork, innovation, attendance, initiative, and the like. Goals for the upcoming year are listed, often weighted by importance or difficulty. There is often a development plan—space for long-term career goals, areas for improvement, and action plans.

Each company has its own variation on this theme. Some companies ask the employee and manager to complete the form separately and then get together to discuss the ratings. Some companies require managers to rank their employees—from top to bottom. (In one high-tech firm, this ranking and rating is known by managers as "ranting and raving.") Not all companies discuss compensation in the same meeting. But basically, most companies follow the same general process. If everyone is doing the same thing, how can it be wrong?

HOW TRADITIONAL APPRAISALS INHIBIT QUALITY AND TEAMWORK

Traditional appraisals work against quality in many ways. In this section, we will discuss five common problems.

Pitting Individuals against Each Other

By focusing on individual performance, traditional appraisals encourage competition. The effect can best be described by a comment from an employee of a Fortune 500 company:

> This teamwork stuff is really a crock. No way will I help another co-worker look good. 'Cause when it comes down to performance appraisal time, I'm being compared to my co-workers. I wouldn't *actively* sabotage his project or performance—I'm not that kind of guy—but help him look good? No way!

This was a dedicated, loyal, long-term employee. This was *not* a problem with the employee. This was a problem with the *system.*

TQM is predicated on an atmosphere of teamwork, but traditional appraisals punish those who seek to help others. These negative consequences are magnified when an organization forces ratings into a bell curve.

Forcing Ratings into a Bell Curve

Many organizations force overall performance ratings into a bell curve to match the curve expected for merit increases. However, this approach encourages competition, rather than cooperation.

The bell curve is intended to discourage managers from rating everyone at the top of the scale. While one might expect this variability within a large organization, significant aberrations occur when this is forced down to a work-group level. Managers with five employees may be required to rate one above average and one below, regardless of performance. This does not encourage the entire team to improve performance, since it compares team members to one another. The comparison that should be made is not between individuals on the team. Rather, the comparison should be made against real competitors or against customer requirements, or both.

Reinforcing the Upright Pyramid

In traditional appraisals, the manager sets or at least approves the goals, monitors performance, and rates the employee. This clearly puts the employee in service of the manager, not the other way around. However, customer-focused organizations invert the pyramid, viewing customers and front-line employees at the top. Managers in this model are expected to support employees. In a traditional appraisal process, even empowering managers find themselves awkwardly forced out of their role of coach and into the role of judge.

Constraining Sources of Feedback

Ironically, managers often are not in a good position to provide feedback to employees. In their daily work, managers spend less time with employees than do the employee's peers, and the peers often know more about the technical details of the work. Also, employees generally spend more time with customers. Yet the traditional appraisal process is an isolated meeting between manager and employee. In those rare cases when peer or customer input is solicited, it is filtered through the manager.

Aggravating Inflexibility

Performance appraisals are tightly structured. They are performed at a certain time of year (usually on a focal review date or employee hire date), use a preprinted form, follow specific procedures, and often are tied to merit increases. The real world of work, however, is not so tightly structured. Annual goals change before the year is up. Projects take more or less than a year. The preprinted form tends to drive the discussion when other questions might be more relevant to discuss. And the value of an employee's contributions may not match the table of allowable merit increases.

WHAT SOME FORWARD-THINKING ORGANIZATIONS HAVE TRIED

In adapting their performance appraisal systems, many organizations have made headway but caused new problems. For instance, at a team-based consumer products plant, self-directed teams were required to do performance appraisals on their peers. However, the focus was on ranking the employees on a variety of criteria. Within a team of 16, someone became number 1 and someone became number 16. This led to rancorous arguments, dividing, rather than uniting, the team.

One Malcolm Baldrige winner designed customer input into its appraisal system. The procedure called for the manager to interview several customers for each employee (usually internal customers within the division) and fold the input into the appraisal. While the concept is a good one, in practice the system overburdened the managers, because they now had to conduct two or more meetings with customers for each employee before conducting the employee appraisals. Even within traditional systems, managers roll their eyes when their human resources department tells them they should meet quarterly with all their employees. The impact on a manager of 15 people (a conservative span of control nowadays) goes something like this:

> 15 meetings × (2 hours to prepare + 2 hours to conduct the meeting + 1 hour to document the results) × 4 quarters in the year = 300 hours per year. This translates into 15% of their time.

The additional requirement of customer meetings swamped the system.

A major chemical company wanted to ensure that employees could give feedback to their managers via an upward appraisal. This took the form of "skip-level feedback," where the boss's boss interviewed the employees. Employees were expected to confess sensitive issues to someone even more intimidating than their own manager, knowing full well that they would still have to deal with the consequences of "tattling." This approach discouraged employees from communicating directly with their managers, from resolving issues head on. And as any child playing telephone knows, the message can get incredibly distorted moving from one mouth to another.

PRINCIPLES FOR REDESIGNING YOUR SYSTEM

These examples embody useful concepts. We should encourage peers to give feedback; we should provide a mechanism for incorporating customer feedback; managers should get feedback from their employees. These components, however, should be embedded in such a way that flexibility, efficiency, and communication are enhanced.

We believe effective performance management systems should be predicated on these principles:

- Merge planning and feedback into a team meeting.
- Design the meeting around thought-provoking questions.
- Align the timing of the appraisal with the work.
- Eliminate individual ratings.
- Provide individual and team feedback.
- Incorporate data about customers and competitors.
- Design a natural way to document results.

These principles form the cornerstone for an "open appraisal," a team-based appraisal process built around an open-ended meeting conducted in an open setting using open (as opposed to closed) questions. [2]

Merge Planning and Feedback

Managers and employees bristle at performance appraisals because the traditional process is laborious and disembodied from the core work. Instead of being tacked on as something extra managers do, it should be integrated into how work gets done.

The solution is to merge business planning and performance appraisal into a single set of meetings. The entire work group should sit down regularly to reflect on the past and to plan for the future. Even in the most autocratic organizations, managers should brief their employees on progress toward goals and results.

This approach transforms the meeting into an interactive team-building process while saving managers considerable time. Instead of conducting multiple individual meetings, managers hold one team meeting. Rather than a stilted meeting between manager and employee, these sessions resemble a regular team meeting in which team members reflect on their recent performance and make plans for the future.

These meetings can form the basis for business plans—rolling up from the bottom instead of being shoved down from the top. Many organizations encourage employees to give input into planning. However, this "encouragement" often is interpreted as perfunctory—a half-hearted attempt by managers to appear interested without relinquishing control. However, inviting employees to participate in the formal business planning process by writing their own business plans can lead to a tremendous wave of innovative ideas. (See Chapter Twelve for more information about modifying the business planning system.)

Design the Meeting around Thought-Provoking Questions

The meetings should revolve around open-ended questions that inspire the employees to think critically about their work and to take ownership for their mini-enterprise. Often these questions fall into four sections:

- *Past performance:* What have been our accomplishments and problems?
- *Future:* What should our goals be?
- *Developmental needs:* What must we improve to meet our goals?
- *Support needs:* What support do we need from others to succeed?

A comprehensive list of sample questions can be found in "Tips, Tools, and Techniques" at the end of this chapter.

Many effective managers recognize the power of asking thought-provoking questions as part of the goal-setting and appraisal process. For instance, Jack Welch, CEO of General Electric, asked five key questions of his managers:

1. What are your market dynamics globally today and where are they going in the next several years?
2. What actions have your competitors taken in the last three years to upset those global dynamics?
3. What have you done in the last three years to affect those dynamics?

4. What are the most dangerous things your competitors could do in the next three years to upset those dynamics?
5. What are the most effective things you could do to bring your desired impact on these dynamics? [3]

While chairman of Chrysler, Lee Iaccoca used a quarterly appraisal process in which he asked his reports three questions: What are your objectives for the next 90 days? What are your plans, your priorities, your hopes? And how do you intend to go about achieving them? Iaccoca comments:

> The quarterly review system sounds almost too simple—except that it works. And it works for several reasons. First, it allows a man to be his own boss and to set his own goals. Second, it makes him more productive and gets him motivated on his own. Third, it helps new ideas bubble up to the top. . . . I've never found a better way to stimulate fresh approaches to problem solving. [4]

The open appraisal simply pushes this process down to the employee level and makes it a team event.

Align the Timing of the Appraisal with the Work

The frequency of the meetings should flow naturally from the work. For many organizations, aligning these sessions with fiscal periods works well. In these cases, a team might hold an extensive annual meeting to plan next year's goal and then meet quarterly to review performance and adjust the goals. Other work groups may need a different schedule. Many jobs revolve around project work. In these cases, the meetings should be scheduled around milestones, with an extensive pre-project planning session and a post-project evaluation meeting. Other industries, such as tourism, may find it more appropriate to align their meetings to the seasons. Each work group should have the flexibility to adjust the timing of their meetings.

Eliminate Individual Ratings

Abolish all forms of individual ranking or rating including a 1–5 scale or such terms as *competent, needs improvement,* and the like. Why? Because rating and berating are one and the same. Even if you resort to a three-tier rating system (which could be categorized as super stars, competent, and about-to-be-fired) as Edward Lawler, the well-known author, recommends, [5] most employees will be forced into what they interpret as "average," which is more likely to make them angry than desirous of improving performance.

You're probably wondering what your corporate attorney would say if you eliminated your rating system. After all, how can an organization fire someone without poor ratings? In fact, ratings have always been a mixed blessing. Frequently, poor performers received generous ratings from cowardly managers that *inhibited* organizations from terminating employees. Certainly if our judicial system can electrocute someone without a rating system and if Nordstrom, a retailer known for excellent service, can operate with an employee handbook printed on a single page, we should be able to justify a termination without putting a number at the bottom of a performance appraisal form.

In any case, the rating is not what justifies termination. Documentation about incidents and behavior has been required to justify the ratings. Organizations can get the necessary documentation related to individual behavior without rating employees. As a protective measure, you may want to implement a peer review process, whereby individuals who feel they have been unfairly treated can have their complaint reviewed by a jury of their peers.

Provide Individual and Team Feedback

Employees should receive feedback from multiple sources, not just the manager. In small teams, each person (including the manager) can receive feedback from all team members and, where appropriate, from customers. Larger teams (with more than eight people) sometimes devise ways to limit the time and number of sources of individual feedback so that the recipient isn't overwhelmed with input and the meeting doesn't drag on. In any case, these guidelines ensure a productive discussion:

1. Let the recipients of the feedback evaluate their own performance first. They should be encouraged to discuss strengths as well as areas for improvement. They should identify one focus area to work on for the next period.
2. After a recipient evaluates himself, his team members add any number of comments about the recipient's accomplishments or strengths.
3. Team members can either agree with the focus item or suggest one other. If they suggest another, they should be able to provide a specific example of how to improve.
4. The manager generally should receive his or her feedback last, so employees who raise sensitive issues do not fear retribution when it is their turn to receive feedback.

In addition to individual feedback, time also should be devoted to discussing how well the team as a whole is operating. Many factors can impact team effectiveness. Trust and openness are critical, of course, but many other factors can cause conflict or poor performance. Research on a variety of team types has revealed eight characteristics of high-performance teams. [6] These characteristics can serve as useful discussion points:

- Clear, elevating goals.
- Results-driven structure.
- Competent team members.
- Unified commitment.
- Collaborative climate.
- Standards of excellence.
- External support and recognition.
- Principled leadership.

Incorporate Data about Customers and Competitors

The appraisal should compare the team's performance to the needs of the organization. Feedback from customers and about competitors is critical to the relevancy of the review. In some cases, it even may be possible for customers to participate in the review process, as is often the case in project work. Other options include using benchmarking data, customer surveys and focus groups, and quality measurement systems.

The team should have ongoing information systems for gathering customer and competitor information and use this information as a basis for planning. For instance, one state highway division on the West Coast leaves customer surveys on windshields and hands them out at truck weigh stations, folding the resulting data into a sophisticated team-based performance measurement system. Its system also provides a basis for comparison against the performance of private contractors.

Design a Natural Way to Document Results

In a TQM environment, teams track their own performance measures on a regular basis. These measures should be reviewed in the open appraisal during the first portion of the meeting, which focuses on past performance.

As a work group refines the open appraisal process, a functional format will emerge. For instance, at Axis Performance Advisors, we refer to our 8-P report: presentations, publications, promotions, prospects, proposals, projects, products, and profits. This structure, which embodies our corporate strategy, forms the agenda in our open appraisal meeting and serves as a basis for documentation throughout the year.

Individual feedback can be documented as part of the minutes of the meeting or be embedded in the process used to facilitate the individual feedback. In one organization, for instance, we designed the feedback process around index cards. All team members documented their own strengths and areas for improvement on the cards and then took notes as others shared their feedback. These index cards were then filed away until the next review meeting.

TIPS FROM A VETERAN

Lee Hebert, plant manager for Monsanto, has embraced the open appraisal concept for his team-based plant in Pensacola, Florida. Monsanto has discovered that sessions of three to four hours is optimal. If the team is not finished, the members schedule another session to avoid burnout. They use the same questions each time. One of Lee's favorites is, "If this were your business, what would you do?" because it builds ownership.

Here is how Lee describes the benefits:

> The open appraisal process is a great vehicle for opening communication with the team. Position power is tough. People tend to tell you what they think you want to hear. So you have to be careful not to use your power to influence the direction of the discussion. But the open appraisal process gets the real issues on the table. Then we can see the possibilities and team members begin to see true growth opportunities.

Lee also notes that open appraisals do not entirely eliminate the need for one-on-one sessions. If performance problems are raised during the meeting, he often follows up with individual coaching. His notes during the meeting serve as documentation. No individual ratings are given, which he feels helps employees see this as developmental opportunity, rather than as a report card. While he feels this process is far superior to traditional appraisals, he cautions that it has not yet been tested in the courts. As a trail blazer, he is willing to take the risk.

CONCLUSION

Based on old paradigms, traditional performance appraisals discourage teamwork and continuous improvement. However, eliminating the function is a mistake. Instead, the process must be reinvented. In this chapter, we have explained how traditional appraisals inhibit quality and teamwork, and we then provided the principles on which new systems should be created.

Tips, Tools, and Techniques
QUESTIONING FOR SUCCESS: SAMPLE QUESTIONS FOR OPEN APPRAISALS

What questions could you include in an open appraisal? Here are some suggestions for each section. The purpose of the questions should not be to get people to commit to action (i.e., What are your goals for the next quarter?) but rather to think about how to improve performance. The group's goals will emanate from the open discussion of their creative ideas. As you generate your list of questions, be selective. You will have a more productive discussion if it is focused around four or five broad questions than if you create the equivalent of a Scholastic Aptitude Test.

Past Performance: Accomplishments and Problems

- What are you most proud about accomplishing in the last period?
- What new skills or knowledge did you acquire?
- How did we improve as a team?
- Did you have any particularly significant personal achievements?
- What persistent problems did we encounter and what were the root causes?
- What do you most appreciate about each of your team members and what one thing would you like each to work on?

Future: Opportunities and Goals

Breakthrough or hoshin [7] questions:

- If you owned this company, what would we be doing differently?

- What is the most exciting concept/technique/process/product that you have heard about outside our organization? How could we use it here?
- Imagine you just discovered that our best competitor cut its cycle time in half. How did they do it?
- From your perspective, what are the three most common causes of quality problems? If you had total control, what would you do about them?
- If we had to bring a product to market in half the time, what would we start doing, stop doing, do differently?
- If a UFO dropped off a group of sophisticated aliens in our organization, what would amaze them? What advice would they give us?
- Imagine that you are overhearing a group of our customers talking at a conference about our new (as yet undeveloped) product/service. They are thrilled! It's just what they have waited for. They can't believe they didn't think of it themselves. It's fabulous! What is it?

Product- or service-related:

- What do our customers tell you they love and dislike about our product/service and what do you think we should do about it?
- We need to reduce the cost of production by Y percent to beat our competition. How would you suggest we accomplish this?
- We need to complete all projects on time and with an effective billing rate of $Y. Describe the status of your current projects and your plan for improving our control mechanisms.
- Where do we lose control over our process and what could be done to get it under control?

Developmental Needs: Areas for Improvement

- If you were able to free up 10 hours a week to do whatever you wanted, what would you do with the time that also would add value to our organization?
- What skills/knowledge will we need on our team in three years? Should we develop those skills or hire people with them?
- Each of us needs to become more versatile. What new products/services/processes do you hope to become qualified to perform? How do you suggest you gain this knowledge?
- Each person on staff is responsible for reading two relevant books a year and reporting on them. What books would you like to read?

- List the skills you would like to improve and state how you intend to improve each skill. Consider the skills needed not only for the work we have performed in the past but also for the future work we may perform.
- How could we work better as a team?
- List any other development goals.

Support: Requirements for Success

- What support have I (your leader/supervisor/coach) provided that you would like to ensure I continue? What should I do more of/less of?
- What can I (or others within the company) do to help you become more productive?
- We depend on a number of other work groups. If we could improve our relationship with and understanding of two other work groups, which groups would you pick?
- Is there something you need to be more productive (new equipment, better maintenance, new procedure, job aid, and so on) and how could we cost-justify the expense?
- What could be done to increase your satisfaction on the job?
- Are there any systems or procedures that inhibit our success?

SUGGESTED READING

Boyette, J., and H. Conn. *Maximum Performance Management: How to Manage and Compensate People to Meet World Competition.* Macomb, IL: Glenbridge Publishing, Ltd., 1988.

Commission Staff on Behavioral & Social Sciences & Education: National Research Council. *Performance Assessment for the Workplace,* vols. I & II. National Academic Press, 1991.

Hitchcock, Darcy. "Performance Management of Teams—A Better Way." *Journal for Quality and Participation,* September 1990, pp. 52–57.

Kohn, Alfie. *No Contest: The Case against Competition.* Boston: Houghton Mifflin, 1986.

Leeds, Dorothy. *Smart Questions: A New Strategy for Successful Managers.* New York: McGraw-Hill, 1987.

Chapter Eight

Rewards

How to Compensate Executives and Employees

Nearly every organization in America today claims to be strongly committed to customer satisfaction and quality, and most of them have embarked on massive quality improvement efforts intended to significantly change their cultures. However, many organizations, even those with progressive approaches to total quality management, continue to pay employees for bottom-line financial performance.

To effectively support their quality effort, organizations need to implement an employee compensation system that strongly links quality and customer satisfaction with pay. In this chapter, we'll discuss several of these nontraditional compensation systems from the standpoint of their strengths and weaknesses. We'll also look at some of the systems that organizations are using to successfully reinforce positive performance with pay, and we'll conclude with a four-step plan to strengthen a weak compensation system.

INADEQUACIES IN TRADITIONAL COMPENSATION SYSTEMS

According to compensation expert Edward Lawler III, pay systems in most American companies have changed very little since the 1950s. [1] Typically, each position has a compensation range that is based on an evaluation of the job and pay scales for similar positions in similar organizations. The salary an individual receives falls within the range established for his or her job grade, with such considerations as length of time with the company, salary agreed on when the position was accepted, and ability to meet annual goals and objectives factored in.

The salary range for most job grades is narrow. For example, pay for a junior engineer might range from $31,000 to $36,200. Raises, usually dispensed once a year, typically range from 3 to 7 percent depending on economic factors and on the organization's financial performance. (Exceptional performers receive the 6 to 7 percent raises, while mediocre performers receive 3 to 4 percent raises.) Usually, all employees, with the exception of senior executives, receive the same benefits and pension plan.

The major disadvantage of traditional compensation systems like the one described above is that outstanding performers earn the same, or just slightly more, than average performers. This type of compensation system does little or nothing to motivate employees to perform at peak levels.

Pay for Performance

In the early 1980s, executives in human resource functions began to investigate alternative compensation systems. Books like *In Search of Excellence* by Peters and Waterman (1982) pointed out that some of the best companies tied employee compensation to performance. "Pay for performance" became one of the hot phrases at human resource conferences, and it appears there is a trend toward more pay for performance compensation systems.

The problems with this approach to compensation are twofold. First, *pay for performance* is a broad term with numerous interpretations—and misinterpretations. Some organizations construe pay for performance to mean top executives receive a profit-sharing bonus. The other problem with performance-based compensation is that, most of the time, the performance paid for has very little or nothing to do with quality and customer satisfaction.

Some of the consequences of poorly designed pay for performance systems are:

- Employees who used to be cooperative and team oriented become fiercely competitive.
- Individual employees become discouraged because all team members receive the same bonus, even when some members did more work than others.
- Individuals and teams become discouraged when they exceed their own goals but receive no bonus, because the company or business unit did poorly.

- Performance does not improve because employees perceive the dollar value of the incentive is not enough to warrant the effort required to earn it.
- Employees figure out loopholes in the incentive system and earn healthy bonuses by beating the system. This leads to changes in the system that close the loopholes which, in turn, discourages employees by making the incentive too hard to earn. For example, account executives who are compensated for generating proposals might inappropriately create proposals just to meet the goal.

DEFICIENCIES IN STRATEGIES THAT LINK COMPENSATION TO TQM

Many of the approaches large organizations are introducing to link compensation with total quality management do not effectively establish the relationship between quality and customer satisfaction and pay. Other strategies result in payment for skills or performance that do not necessarily support quality or customer satisfaction objectives, and some reinforce inappropriate behavior. Six of the most common of these strategies are:

- Linking quality to merit increases.
- Linking quality to executive bonuses.
- Pay for skills.
- Profit-sharing bonuses for all employees.
- Guaranteed contracts for executives.
- Gainsharing.

In the following discussion, we'll address each of these strategies, along with their inherent deficiencies.

Linking Quality to Merit Increases

Relating merit increases to quality achievement is the revealing way for organizations to link compensation and quality. With this approach, once a year an employee and his or her manager agree on specific quality and nonquality related objectives or goals. If the employee achieves the goals, he or she receives a merit raise.

Despite the popularity of merit increases, they do not motivate the long-term behavior changes required to achieve quality and customer satisfaction. One reason is that the increases are too small. In most organizations, merit increases range from 3 to 7 percent of annual salaries. Typically, the raises outstanding performers receive are only 1 or 2 percent bigger than the raises the average performers receive. For most employees, this increase is not significant enough to motivate superior performance over an extended time.

Another limiting factor of merit raises is that they are given annually, which is not often enough to influence behavior throughout the year. Also, the raise loses much of its impact because it is based on a variety of factors such as seniority, performance, and attitude. Merit raises also do nothing to encourage behavior changes because they reward only past performance. The raise an individual has already received is never at risk.

Linking Quality to Executive Bonuses

Unlike merit increases for employees, bonuses ranging from 20 percent up to 200 percent of salary are usually a major portion of an executive's total compensation. This approach also fails to achieve its objective, because the weight given to quality and customer satisfaction in the overall determination of the bonus is often insignificant. In 1992, one large manufacturing company related the bonus to the score an executive's organization received on an internal assessment. Although the assessment was based on the Baldrige Award criteria, in the final translation the quality score was worth only 5 percent of the total score. Ninety-five percent of the bonus was based on financial and operational measures of performance.

An additional problem with the practice of linking quality to executive bonuses is that, in most organizations, only a handful of people are eligible. Bonuses usually are restricted to the top two or three layers of management. In a company of 5,000 employees, this may mean 50 people. Although this approach might inspire a strong commitment to quality among this small group of managers, it does nothing for the majority of employees.

Pay for Skills

Another major trend gaining popularity is to pay people based on their skills and knowledge. With this approach, an employee receives a higher salary or hourly wage based on the number and types of skills he or she masters. Employees who want to increase earnings can learn additional skills and demonstrate their mastery through paper-and-pencil or performance tests. The organization only pays more if the employee learns more valuable skills and becomes a more valuable employee. Employees benefit, because they can choose to increase pay by mastering new skills.

This approach to compensation can be fair to both the organization and the employee. At ARCO Alaska, for example, the technicians who maintain the oil pipeline and pumping stations at Prudhoe Bay are on a pay for skills system. They can earn more money by learning and demonstrating mastery of new skills on performance tests that require actually performing the job tasks to a standard. ARCO contends that the company benefits by having a more knowledgeable and flexible workforce, paying employees more only when they are worth more to ARCO. Employees like the system, because they can choose to increase pay. Many say their jobs are more enjoyable, because they are not limited to doing the same tasks year after year.

Problems with skill-based pay. The popularity of skill-based pay systems is growing. A 1987 study by the American Productivity and Quality Center revealed that 8 percent of manufacturing companies have such plans and that 2 percent of service companies have knowledge-based pay plans in place. However, there are disadvantages to pay for skills systems. The major one is that, while this approach rewards training and the acquisition of new skills, it does nothing to promote the improvement of quality and customer satisfaction. Skill acquisition should lead to improved job performance, but often it does not. A good compensation system should reward accomplishments that are tied to the organization's goals.

Another problem with skill-based pay is explained by Les Schroeder, Pacific Bell compensation manager, in the September 1990 issue of *Training Magazine:*

> We have chosen not to have skill-based pay because it pays for skills whether a person actually in a job is using those skills or not. [2]

Increasing employees' pay increases an organization's costs and decreases profits, unless it can be demonstrated that increased expense results in financial benefits to the organization. The organization benefits only when employee knowledge and skills are applied on the job.

Profit-Sharing Bonuses for All Employees

For decades, organizations have used profit sharing to motivate executives to keep their eyes on the bottom line. Recently, these plans have filtered down to the working level in big companies. This approach to compensation is based on the concept that if all employees are eligible for a profit-sharing bonus they will be more conscientious, work harder, and look for ways to cut costs and improve the company's profits. Often this plan is credited with enhancing teamwork, as well, since it encourages employees to work together to make the company more successful.

Profits are a major output for the organization, and paying employees for profits is preferable to paying them for skills. This approach also is fair. Employees share when there are profits and receive no extra money when there are no profits. The biggest difficulty with profit sharing is that, like pay for skills, it does not encourage improvements in quality and customer satisfaction. Profits are related to quality and customer satisfaction; however, many factors besides quality influence profits. The profits of some of the big oil companies, for example, have little to do with levels of quality or customer satisfaction. Some years, oil company executives receive huge bonuses because oil prices are up. Other years, they may work harder and receive little or no bonus because lower oil prices have diminished company profits.

Another problem with profit sharing is that most employees have very little control over the profitability of the corporation. Although every employee's performance can impact the bottom line, the individual's contribution is too far removed. Frequently, an individual, a department, even an entire business unit performs at exemplary levels but receives no bonus, because the whole company did not make an acceptable profit that year.

Guaranteed Contracts for Executives

A guaranteed employment contract has become a common way to lure an executive from one company to another. Usually, there is a great deal of economic security for the executive. A typical contract might guarantee a base salary of a given amount, say $220,000 each year, for a specified period, perhaps five years. In addition, the contract usually stipulates that the executive cannot be fired or laid off for performance or economic reasons. Although it appears that all risks are on the company's side, companies argue this type of contract is necessary to secure the talent they want, as well as to convince the executive to leave her or his current position and company.

The guaranteed contract arrangement resembles the contract practices in professional sports, and, indeed, the two have much in common. The motivations of the owners and the talent are similar to those of companies and executives. In fact, some attorneys who handle salary negotiations for athletes now also are handling executives.

Guaranteed contracts for executives are pursued by companies because they don't want to invest in an individual who may leave in a year or two for a better offer. The company wants to keep the individual until it gets a return on its investment in his or her talents. Contracts appear to be a sound approach for both the company and the executive. However, the guarantee doesn't serve an organization's long-term interests. Guaranteed contracts for executives are one of the most damaging things a company can do if it wants to improve quality and customer satisfaction.

Richard O'Brien and his associates at Hofstra University studied the performance of major league baseball pitchers over a six-year period during which they had long-term guaranteed contracts. The results of this study, which appear in the book *Industrial Behavior Modification* (1978), reveal the following:

> Major league baseball pitchers' performance declines on all key measures (earned run average, wins, strikeouts, hits given up, etc.) once pitchers are put on multi-year mega-buck contracts. During the same time period, the pitchers who were not on long-term contracts improved their performance. [3]

The baseball pitchers with guaranteed contracts usually received compensation that was three to four times higher than their counterparts who did not have guaranteed multi-year contracts. These pitchers had annual contracts, and their compensation and contract renewal were based on performance each season.

A similar phenomenon has been said to occur after university professors receive tenure. Research, publications, and teaching excellence often decline following tenure. Here, and on the ball field, individuals are being allowed to relax. They no longer need to hustle to keep their performance at peak levels.

Tracking executives' ERA. What happens to baseball pitchers and professors probably happens to executives with guaranteed contracts as well. However, tracking an executive's earned run average may not be as simple as tracking a major league baseball pitcher's. The only performance these contracts guarantee is that the executive won't leave within a specified number of years. Companies that imagine these contracts guarantee a return on their investment in the executive are mistaken. High salaries do nothing to promote good results. American executives are among the highest paid in the world, yet American products and services still show much room for improvement in quality. Even when executives can earn a $100,000 bonus for good performance, the security of the guaranteed contract appears to have a bigger influence on determining his or her level of performance and risk taking.

Gainsharing

Gainsharing is an old approach to compensation, dressed up with a new name. The concept has been around for more than 50 years, under such names as the Rucker Plan, the Scanlon Plan, and Improshare. Since the inception of the total quality management movement, gainsharing has been repackaged as an alternative compensation plan designed for the employees of the 1990s.

Gainsharing is more a philosophy than a specific compensation system or plan. The philosophy of gainsharing is well described in the Fall/Winter 1991 issue of *Tapping the Network Journal* by consultant Robert Masternak:

> It is best described as a system of management in which an organization seeks higher levels of performance through the involvement and participation of its people. As performance improves, employees share financially in the gains.

From this description, it is obvious why the TQM movement has become so enamored of gainsharing. It fits perfectly with quality improvement teams, self-directed work teams, and the concept of empowerment. Gainsharing often involves awarding cash bonuses to employees either monthly or quarterly, based on the ability of teams to improve specific aspects of performance within their own job responsibilities.

Why Does Gainsharing Fail?

While the concept basically is sound and can be extremely valuable in facilitating the implementation of TQM, most companies have not implemented gainsharing correctly. Some of the problems include:

- Gainsharing plans are usually designed by copying (benchmarking) other company's plans or by acquiring a packaged approach/plan.
- The dollar amount of gainsharing bonuses is often not perceived as enough to compensate employees for the work required to earn the bonus.
- Bonus payments are based entirely on team performance, even if all individuals on the team did not contribute equally to earning the bonus.
- Payments are often tied to performance measures over which employees have little control, which leads to frustration.

How to Make Gainsharing Work

Of the alternative compensation approaches we've discussed in this chapter, gainsharing has the most potential for success in a company trying to implement total quality management. Because of this potential, it's worth reviewing some simple rules that will help make gainsharing more effective.

1. Design the plan with the input of the customers/users of the system—the employees. Control Data has had success with gainsharing. In the December 1991 issue of the *Journal for Quality and Participation,* Robertson and Osuorah explain their system was designed by teams that included employees from various levels. [4] Control Data studied the successful gainsharing approaches used at both 3M and Xerox but did not copy these approaches. Rather, it took the best practices from both companies and adapted its own approach.

2. Make sure that the dollar amount of the bonus is large enough and frequent enough to motivate employees. A large manufacturing company had thousands of quality improvement teams working on projects, which resulted in changes or new processes that dramatically improved quality and saved the company thousands of dollars. As part of the gainsharing program, employees on these teams received $50 savings bonds and a certificate. However, what the company intended to be a nice "thank you" carried negative connotations for employees. Many resented what they viewed as an insignificant acknowledgement of their considerable effort. Perhaps some type of symbolic award that provided employees with recognition from their peers, or a cash bonus based on a percentage of the savings the teams produced, might have been a more successful motivator.

3. Base payments on individual and team performance. A large insurance company implemented a gainsharing plan based entirely on measures of team performance. Teams that solved problems or improved processes received cash awards, which were based on a percentage of the value of their projects. Feedback from employees revealed that many did not think the gainsharing plan was fair. Individuals on teams usually contribute unequally to the gains achieved. If everyone on the team receives the same reward, those that did the majority of the work resent those who contributed less.

Another company distributes gainsharing awards to both teams and individuals. This approach is considered more equitable by employees, and it allows the company to apply gainsharing to situations where teams are inappropriate.

A gainsharing plan might reward individual employees for the suggestions or ideas they generate and implement to improve company performance. In the same company, there may also be a gainsharing plan that rewards teams for projects completed that produce benefits for the company.

4. Make sure that gainsharing bonuses are based on performance indexes that employees can influence or control. The best gainsharing systems are based on measures that are tailored to each department or function. One organization had an excellent approach to gainsharing: At the start of a team project, it established the measures of success and the reward that would be distributed. A committee of employees from various levels and functions reviewed proposals that teams presented for improvement projects. The committee determined the measures and incentives before a team was given approval to begin the project. This provided flexibility for customizing it to each individual improvement project.

COMPENSATION SYSTEMS THAT WORK

Many organizations have implemented nontraditional compensation systems that successfully integrate quality and customer satisfaction with employee compensation. We will address many of them in the following discussion.

Incentive-Based Pay for All Employees

One effective compensation system is to put all employees on incentive-based pay, with a large percentage of the incentive being based on measures of quality and customer satisfaction. For example, at Federal Express, every employee from the truck driver up to CEO Fred Smith is evaluated on three dimensions:

- People.
- Service.
- Profit.

The "People" component of the system is a rating that each person is given by subordinates or peers on their performance as a supervisor, manager, or team player. Since employee satisfaction is considered a prerequisite to the achievement of customer satisfaction, employees evaluate their managers and supervisors on managerial skills. This score is a requirement for earning bonuses, which indicates how serious Federal Express is about employee satisfaction and good leadership skills.

The "Service" component is the customer satisfaction rating. Employees who do not deal directly with external customers receive the rating based on internal customer satisfaction. In many cases, employees serve several different customers. For example, the corporate finance department might list the following groups as its primary customers:

- Field finance people.
- Corporate executives.
- Employees who use the department's services and products.
- Outside regulatory agencies.
- Shareholders.

Employees who fail to achieve the standard for either the people or service dimensions of the formula do not receive a bonus that quarter. For instance, if employees consider their manager exemplary, but customers think he or she does not care about customer interests, the manager does not receive a bonus. Conversely, if a manager does everything possible to keep customers happy, but employees give him or her low ratings for leadership skills, again the manager would not receive a bonus.

The amount of the bonus for employees who meet the standard in both people and profit is determined by their performance in financial measures. Ratings for the three factors are computed for each employee once each quarter, and bonus checks are paid with the same frequency.

The overriding advantage of the Federal Express system is its simplicity. Every employee is measured on the same three types of variables, which are tailored to each employee's individual function and job. Bonuses are paid quarterly based on the people–service–profit scores.

The contributing factor in the system's success is not the bonus money—typically, bonuses are small—but the quarterly feedback that each employee receives. The feedback and the bonus help employees at all levels focus on what is important: satisfied employees, satisfied customers, and a profitable company.

Performance-Based Compensation

Another innovative employee bonus system is used by Australian and New Zealand Direct Lines, a Long Beach, California, company that ships cargo to Australia. In this system, every employee is evaluated on three major measures of performance. Customer satisfaction is required to be one of the three. The other two are operational measures, such as meeting deadlines, budget performance, or meeting financial targets. In some cases, the measures are based entirely on individual performance. In others, they are based on team performance. Employees can track performance against goals, and they can earn bonuses of several thousand dollars based on their performance on the three key measures.

CEO Michael Beard convinced the parent company to make the investment in the bonuses, believing that the investment would pay off. So far, it has paid off extremely well. Not only are sales and profits up, but the company's market share has increased due to high customer satisfaction scores.

Bonuses Tied to Customer Satisfaction

Coldwell Banker Relocation Services is a small business unit within the Sears network. The organization provides relocation services to major corporations, such as IBM, GTE, and others. Coldwell Banker Relocation Services went from sixth or seventh in market share to number two by concentrating on consistent, excellent service.

Coldwell relocation counselors, the personnel working directly with the transferring of executives and their families, receive cash bonuses each quarter based on customer satisfaction ratings. Bonuses for managers and other personnel in the organization also are tied to customer satisfaction scores. During 1992, employees who provide support services to the relocation counselors also were added to the incentive pay plan.

Coldwell Banker Relocation Services has written in goals for customer satisfaction, gains and losses of accounts, and market share as part of its overall strategic plan. Performance against the goals outlined in the plan determines executive bonuses. As a result, everyone, from the president to individual contributors, focuses attention on customer needs.

Pay for Service

Financial service organization Charles Schwab & Company deviates from the pay practices traditionally used in brokerages and financial firms. Schwab pays its people for service, not for transactions, according to a report in *Human Resource Executive.*

Schwab realizes that service is all it sells, so that's what it bases employee rewards on. Bonuses for managers and for customer contact personnel are based solely on surveys designed to measure levels of customer satisfaction. Personnel who work in behind-the-scenes support functions receive bonuses based on levels of internal customer satisfaction. Executives at Schwab believe that delighting customers will give them an edge over competition both now and in years to come. Schwab is one of the few companies willing to bet money on this, however.

Schwab's growth and financial success over the last few years testify to the success of concentrating on service, rather than on short-term financial results. The company also was recognized by the Tom Peters Group for excellence and high levels of customer satisfaction.

Pay at Risk

In pay-at-risk compensation systems, an employee's compensation is derived partly from a base salary and partly from a bonus. Since the bonus is given only if the employee meets specified performance standards, this portion of the pay is at risk. For example, if an employee is paid a salary of $50,000 and is eligible to earn a bonus of up to an additional $25,000, then a third—or 33 percent—of the employee's compensation is at risk. When assessing a company, Baldrige examiners look at the percentage of employee compensation that is at risk.

In his book, *Thriving on Chaos,* Tom Peters explains that improving compensation systems requires the following approach:

> The suggestions are simple: (1) very low base pay compared to very high incentive pay, and (2) rewards based upon what you want to happen. [5]

The pay-at-risk approach to compensation is gaining popularity in many organizations that are serious about TQM. In most organizations, top-level executives have pay at risk. Steelcase, the premier manufacturer of office furniture and fixtures, has about 60 percent of management compensation at risk. However, the trend is to put a significant portion of every employee's pay at risk, basing the amount of bonus or extra compensation, at least partially, on measures of quality and customer satisfaction.

CONCLUSION

An organization's total quality management initiative must be supported with a compensation system that encourages and motivates employees to achieve the desirable performance. Most effective is one that links quality and customer satisfaction with employee pay. Organizations that are serious about achieving quality and customer satisfaction must integrate these aspects of TQM into their compensation system; however, most have failed to make this link. Others link compensation and quality, but the link is weak.

Although several of the nontraditional compensation systems we discussed in this chapter do not effectively reinforce desirable performance, some of the systems we discussed do establish a strong relationship between desirable performance and increased pay. The companies that have implemented these systems are finding improved quality, increased customer satisfaction, and improved job performance to be the ultimate reward.

Tips, Tools, and Techniques
IMPROVING COMPENSATION SYSTEMS WITH FOUR QUICK STEPS

1. Eliminate raises based on cost of living or on seniority.
2. Put every employee on incentive pay.

3. Gradually increase the percentage of pay at risk until it is at least 25 percent.
4. Give quality and customer satisfaction measures significant weight in bonus determination.

SUGGESTED READING

American Productivity and Quality Center. *The Use of Skill-Based Pay in U.S. Companies.* Houston: American Productivity and Quality Center, 1987.

Ehrenfeld, Tom. "The Case of the Unpopular Pay Plan." *Harvard Business Review*, January/February 1992, pp. 14–23.

Hequet, Marc. "Paying for Knowledge in Paper Factories." *Training,* September 1990, pp. 69–77.

Lawler, Edward E. III. *Strategic Pay.* San Francisco: Jossey-Bass, 1990.

Masternak, Robert. "Gainsharing at B. F. Goodrich: Succeeding Together Achieves Rewards." *Tapping the Network Journal,* Fall/Winter 1991, pp. 13–16.

O'Brien, R. M.; A. M. Dickinson; and M. P. Rosow. *Industrial Behavior Modification.* Elmsford, N.Y.: Pergamon Press, 1978.

Peters, Tom. *Thriving on Chaos.* New York: Alfred A. Knopf, 1988.

Peters, T. J., and R. M. Waterman. *In Search of Excellence.* New York: Harper & Row, 1982.

Robertson, R. N., and C. I. Osuorah. "Gainsharing in Action at Control Data." *Journal of Quality and Participation,* December 1991, pp. 28–33.

Sulit, B. K. "Benchmarking HR: Measuring up to the Leaders." *Human Resource Executive,* June 1992, pp. 23–39.

PART III

WHY ORGANIZATIONS FAIL DURING INTEGRATION

For TQM to succeed, it must eventually become integrated into the fabric of the organization, driving organizational beliefs, structures, practices, systems, and organizational learning. When this occurs, TQM becomes invisible, a ubiquitous field that binds the organization together. Such is the case at Federal Express, a winner of the Malcolm Baldrige National Quality Award. According to Baldrige examiners, Federal Epress almost did not qualify as a finalist because TQM was so well integrated that few of Federal's quality efforts could be singled out in isolation.

This integration challenges many organizational bastions, forcing organizations to confront closely held beliefs and change what best-selling author Peter Senge calls their "guiding ideas." For decades, information and power were hoarded by management, but TQM requires that these be shared with employees. This empowerment forces changes to management philosophies and practices. Organizational structures must be dismantled and organizational systems must be aligned to quality.

These dramatic changes threaten many people in executive suites, for, up until "integration," many may have been able to avoid making significant personal changes. However, further progress in TQM is now directly tied to their personal development. After being reinforced and advanced for traditional practices, some executives cannot make the leap to new paradigms. TQM fails in Phase 3 for the five reasons discussed below.

Failing to Transfer True Power to Employees

Quality improvement teams used in Phase 2 rarely threaten management power, because the teams are given only the power to recommend. For an organization to reach high performance, however, employees must be given vastly greater levels of authority. While terms like *self-direction* and *self-managed teams* strike terror in the hearts of many mid-managers, these are required for survival in this knowledge-based economy. Chapter Nine explains how to overcome the obstacles to advanced forms of empowerment.

Maintaining Outmoded Management Practices

Employees are quick to notice inconsistencies in management behavior. Not "walking the talk" is a major cause for failure in both TQM and empowerment efforts. While managers crave simple answers, we cannot prescribe specific behaviors, for slavish attention to techniques will not work. There must be an integrity of management philosophy, practices, and behavior. However, changing belief systems is one of the most challenging tasks. Chapter Ten addresses how to help managers make the transition.

Poor Organization and Job Design

The principles of organization design for the old economy will choke an organization in the new economy. In the past, organizations have oscillated between centralized and decentralized structures, as if any point along the pendulum swing is as good as another. Functional structures with stovepipe communication systems must be dismantled in favor of more fluid networks. Similarly, jobs must be redesigned to allow for the increased power and discretion employees must wield. Chapter Eleven provides the principles by which the organizations and jobs of the future must be designed.

Outdated Business Systems

Some human resource systems became major obstacles in Phase 2. However, now in Phase 3, the remaining business systems must be reinvented. Current accounting practices are inadequate to the needs of a total quality organization. Business planning systems provide no structured methods for those closest to the customer to have input into the business plan. Information systems still focus on managing a tangle of internal data instead of providing competitive advantage. Chapter Twelve reveals leading-edge practices and provides a method for redesigning business systems to support quality.

Failing to Manage Learning and Innovation Diffusion

In most organizations, innovative solutions to quality problems are generated in pockets, but the learning is never assimilated into the rest of the organization. Even less likely to be shared are the significant failures, yet these are of equal value to the advancement of quality. For most, becoming a learning organization is only a vague concept. Chapter Thirteen provides a structured approach to institutionalize organizational learning.

For TQM to transcend being a " program," quality practices must be integrated into every facet of the organization. This transcendent phase carries both the greatest threat and the greatest payback, for it involves reinventing how work gets done, and there are few examples to follow. The quality paradigm with its clear vision and unwavering values must guide us as we step into the future together.

Chapter Nine

Power Structure

How to Empower Employees

Unenlightened organizations try to do TQM *to* their employees instead of *with* them. Through the lens of these traditional organizations, TQM often is seen purely as another management cost-cutting tool. TQM will fail in these organizations because managers are unwilling to share their power with employees. TQM requires that those closest to the work be empowered to correct problems or defects. For TQM to succeed, front-line employees must be given extensive powers.

In this chapter, we will examine why empowerment is critical to TQM and discuss the demonstrated benefits of employee involvement. We will explain how self-directed teams, an advanced form of empowerment, can be used in conjunction with TQM to yield a high-performance work system. We also offer solutions to common obstacles that inhibit their implementation.

WHY MANAGERS HAVE ALL THE POWER

Management as a discipline has been around for less than 100 years. Prior to that time, most work was performed by small bands of workers who, like a string quartet, did not need a conductor. Only in war was it necessary to coordinate the actions of many people. And so our management practices evolved from the military paradigm: officers/managers versus troops/employees, giving orders versus taking, thinking versus doing. Obedience equaled loyalty. Doing what you were told became confused with good performance. Employees were viewed as disposable, faceless resources. Uniforms and protocol preserved a safe distance so power and control could be maintained.

This paradigm is so well rooted that many managers cannot describe their management philosophy and its origins. To them, it's just the way it is. As they become indoctrinated into the culture, they simply accept that their role is to figure out what should be done and to tell people to do it. Worse, many have become addicted to the trappings of position, tenaciously protecting their company cars, mahogany desks, pension plans, and Persian rugs.

This paradigm is ill suited to the demands of today. In an age when competitive wars are won by speed and flexibility, top-down control only gets in the way. Consider the highway maintenance crew confronted with spring flooding across a major state highway. The crew wanted to close the road to divert the water but the members were told that they did not have the power to make that decision. Their supervisor was dispatched to review the situation. Then the supervisor's boss . . . the boss's boss . . . and then the boss's boss's boss. Four days later, after much of the road had washed out completely, the now disgruntled crew was given the go-ahead. Could the risk of empowering the crew to act ever have cost as much as four days of flooding, four layers of management involvement, and four bitter crew members?

Paradoxically, the more managers count on controlling, the more out of control things become; the more managers share their power, the more powerful they become.

FIVE LEVELS OF EMPOWERMENT

Over the years, organizations have tried to empower their employees to varying degrees. According to the best research available, the greater the empowerment, the greater the payoff. [1]

Empowerment can be viewed along a continuum. Over the last 50 years, organizations have experimented with different structures to empower their employees, slowly moving up the empowerment scale (see Figure 9–1). The performance curve is intended to be illustrative, not literal.

Traditional Management

At the bottom of the empowerment scale the military paradigm is in full force, with only the leaders empowered. Even the military now recognizes its faults, reportedly having directed the Gulf War from the front lines and

FIGURE 9–1
Empowerment Scale

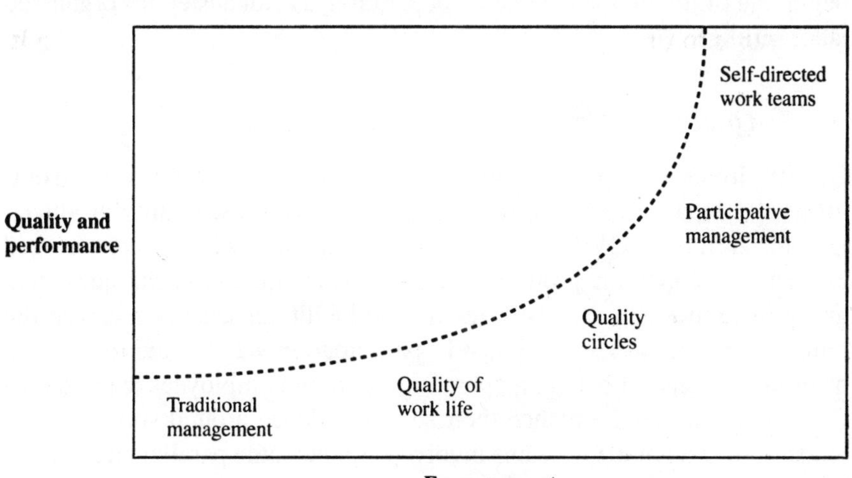

Source: AXIS Performance Advisors. Used with permission.

not from the war room in Washington. The problem with traditional management, however, goes beyond its autocratic style. Traditional organizations usually are organized functionally, putting similar people together into departments: engineering, production, human resources, accounting. This structure encourages fiefdoms and leads to huge walls that separate interdependent parties.

Quality of Work Life

In the 1970s, quality of work life (QWL) teams became popular, especially where employees were represented by unions. These teams were brought together on a regular basis to identify ways of improving the quality of the product or the quality of their work life. Not surprisingly, given the choice, the majority of QWL teams focused on improving their work life. While these teams did improve employee satisfaction and in some cases reduced grievances, few demonstrated improvements in organizational performance. These teams did little to affect the power structure since the QWL teams could only make recommendations. These teams, however, were the beginning of union–management cooperation, a critical step for organized labor settings.

Quality Circles

Quality circles (QCs) became popular about the same time as QWL, an outgrowth of Japan's ability to reverse its image as a low-quality producer. QCs differed from QWL teams primarily in that the whole team received extensive training on group decision making, quality techniques, and group dynamics. (In the QWL teams, the facilitator usually received the bulk of the training.) As with QWL, participation was limited to periodic meetings, instead of being an operating norm; but employees learned critical skills that could enhance their self-confidence around solving problems. Both QWL and QC teams involved cumbersome parallel hierarchies: layers of interconnected committees operating outside the normal chain of command.

A recent study that examined the results of improvement efforts in 131 organizations over 30 years discovered that quality circles and similar team problem-solving structures can have a *negative* impact on financial performance. [2] It seems that the cumbersome committee structures and processes increased costs and bureaucracy overall. However, these two efforts provided many critical problem-solving and group process skills for future participative efforts.

Participative Management

Participative management became acceptable practice following published works on Theory Y and Theory Z management styles. It is defined as managers sharing their power and influence by regularly asking employees for input while maintaining at a minimum a veto over their ideas. This represents a significant increase in empowerment from the previous two types, because employees are expected to participate more often than once a week for an hour.

However, the power structure really has not been altered because managers are still firmly in control. Lee Iaccoca's self-described style provides a perfect example:

> My policy has always been democratic all the way to the point of decision. Then I become the ruthless commander. "Okay, I've heard everybody," I say. "Now here's what we're going to do." [3]

While Iacocca might want to review the definition of democracy, he certainly got results as demonstrated by Chrysler's resurgence.

Self-Directed Work Teams

Self-directed teams represent a significant change in the power structure, because management gives up its veto around clearly defined areas of responsibility. The self-directed team assumes most, if not all, the responsibility of a traditional supervisor. These teams plan, conduct, improve, and evaluate their own work. They may interview potential candidates as well as train and appraise their own members. They also may have direct contact with customers and suppliers. Often, team members are cross-trained, and job classifications are combined, leading to increased flexibility. Some self-directed teams even hire, fire, and compensate their own members.

Empowerment and Performance Are Linked

The empowerment scale depicts the relationship between empowerment and performance. Early experiments with nominal levels of empowerment yielded slight increases in quality and performance. However, when the power structure is significantly altered—as with self-directed teams—performance tends to improve dramatically.

According to Edward Lawler's research, significant improvements in performance generally do not occur until the degree of empowerment offered by self-directed teams. [4] Until that point, true power has not shifted, and involvement can be interpreted as tokenism. But once official responsibilities are delegated, the performance of the organization takes off.

While the work may place limitations on the degree of empowerment that is practical, over 45 years of data support the conclusion that redesigning the workplace around high-performance, self-directed teams yields quantum leaps in performance. They yield significant inprovements in quality, decision making, work methods, employee retention, and safety. [5] Furthermore, they have worked successfully in a variety of settings, including manufacturing, schools, government, health care, and financial services. (See Results of Empowerment at the end of this chapter.)

HOW SELF-DIRECTED TEAMS DIFFER FROM TRADITIONAL WORK GROUPS

For TQM to work, employees must be empowered, and self-directed teams are currently at the extreme end of the empowerment scale, representing the highest degree of empowerment and concurrently the highest improvements in quality. We already have said that self-directed teams assume most or all of the responsibilities of a traditional supervisor. However, this simple definition does not begin to describe how the workplace changes.

For self-direction and TQM to succeed, empowerment must be supported in the following three ways.

Organizational. Organizational structure and systems (such as appraisals and rewards) must support quality and teamwork. Sometimes a functional organization must be dismantled to put interdependent people together (see Chapter Eleven). Job responsibilities and roles must change at all levels.

Interpersonal. People must interact in empowering ways. For instance, managers must learn to help employees solve their own problems, rather than tell them what to do. Relevant knowledge must be valued over status, collaboration over competition.

Individual. Employees at all levels must gain the skills and confidence to take on greater responsibility. They also must internalize the new paradigm and values.

Until all three supports are in place, true empowerment (as embodied by self-direction) cannot occur. The following Hewlett-Packard example demonstrates each of these three components.

Organizational Empowerment

Late in 1990 and into the first half of 1991, Hewlett-Packard in Corvallis, Oregon, redesigned its Think Jet Pen department (which manufactures printer cartridges) into self-directed teams. At an organizational level, team members now are responsible for most supervisory and technical support tasks, including staffing, planning, communicating to internal customers and to some vendors, and peer appraisals. They even purchase their own capital equipment and are procuring automation equipment without engineering support.

How these responsibilities are shared is as important as the responsibilities themselves, for they should be dispersed among team members, not hoarded by one team leader. At H-P, the team has established up to 15 coordinator roles through which team members rotate. These include a financial coordinator, who tracks expenses and targets; a staffing coordinator, who proposes staffing changes to accommodate skyrocketing demand for their product; and a scheduling coordinator, who coordinates with upstream and downstream operations and vendors.

Interpersonal Empowerment

On an interpersonal level, people now interact in new ways. Although many team-based organizations emphasize structured team meetings, problems are solved at H-P on a real-time basis through impromptu meetings. It is not unusual to see an operator dashing down the hall to get engineering assistance; but, instead of handing off the problem as they would have in the past, the operators maintain control of the problem-solving process.

Managers had to delegate most of their traditional functions and become team facilitators. Kathleen Baker, who helped facilitate the team's transformation, describes the relationship between management and employees in this way:

> Now employees see managers not as an elite group, whose purpose is to make them miserable, but, rather, as a group that relies on them for input and acts on it.

Personal Empowerment

On a personal level, these changes in roles and responsibilities add a measure of stress and conflict. Some employees felt a loss of status. The old production operator role, for instance, required that the operators have special knowledge and, with self-direction, these specialists were expected to share that knowledge with everyone through cross-training. As one exproduction operator said, "At first, I didn't want to give anything up. But then I gave up little pieces—I mean *really* little pieces—until I saw they could handle it. Finally, I realized that I couldn't do everything I needed to do if I held onto all these tasks."

How has this affected the employees? Their "transition coach" has a one-word answer: "Ownership. Now team members see themselves as managing their own small business."

None of this was easy. There were times when the team wanted to give up and go back to the old way. But, with all the ups and downs, there has been a tremendous payoff. When asked if the team would ever go back to work in a traditional organization, the members gave a resounding "No!" And the organization has enjoyed a 38 percent increase in pens produced per person on the team since fiscal year 1990 and a 150 percent increase in pens produced for each technical staff member who supports the team (e.g., engineers and technicians). In addition, their variable production expenses per unit have been reduced, and the managers' span of influence has increased from 11:1 to 45:1 or more.

HOW SELF-DIRECTED TEAMS IMPROVE QUALITY

Why is it that front-line employees can succeed when managers cannot? After all, wasn't the camel designed by a committee? Managers and other college-educated professionals receive two-thirds of the U.S. training dollars [6] and usually possess a more sophisticated understanding of the organization as a whole. Why shouldn't they be in a better position to make decisions? In fact, a unilateral decision is much faster to make than one requiring team deliberation. Why not leave employees alone to do their "real work"?

Part of the answer is as simple as this: there are more employees than managers. If only managers are supposed to think about how to improve work, then only a small fraction of an organization's brainpower is being utilized. Even with suggestion systems, employee of the month awards, and participative management practices, a vast majority of the American workforce feels disempowered because they are not given the authority for implementation.

The complete answer, of course, is more complex and subtle. Not only are people more likely to be committed to a decision when they participate in the process, many of their spiritual needs can be met as well. Properly managed, self-directed teams can provide a compelling purpose, a sense of belonging, strong social bonds, and a sense of control. Allowing employees to manage their work and solve their own problems, coupled with a strong sense of vision and purpose, provides a powerful motivator that improves performance.

In an auto manufacturer, a work group griped for years about another work group. The group wanted to create a wider aisle for transporting demolished materials to railroad cars so the members would not damage the fragile inventory on either side of the aisle. For years, the employees had asked their supervisor to fix the problem and, for years, the supervisor had said he had tried, to no avail. After becoming a self-directed team, the workers took matters into their own hands, holding a meeting with the other department. The problem was solved in two weeks.

The solution seemed so simple. Why couldn't the supervisor have solved it? Perhaps he could have; who knows. The principle here, though, is to let those who care most about a problem solve it. And not just *let* them solve it but let them know it is their *official responsibility* to solve it.

Self-directed teams are more productive for other reasons as well. When self-directed teams are designed, they often change the existing organizational structure, eliminating functional fiefdoms. For instance, Northwest Natural Gas in Portland, Oregon, analyzed its work flow only to discover that, to schedule a customer's request for service, the order passed through three departments and eight pairs of hands. This process wasted time and increased opportunities for error. The company redesigned the work around a single self-directed team by pulling individuals from the various departments. Now the team handles order processing from beginning to end. As a result, errors were reduced by 70 percent and the average time to hook up a customer dropped from three days to six hours.

As with Northwest Natural Gas, self-directed teams should be designed around a whole process, product, project, or customer. Completing a whole makes performance measurement easier and increases the workers' identification with the output and its quality. For instance, Volvo did what most U.S. auto manufacturers said was impossible: it so modified its process that a team assembles an entire car. Wouldn't you feel more committed to quality in that setting than if your job was to tighten lug nuts all day long?

Also, team members often are cross-trained on each other's jobs. This cross-training provides two benefits. First, this cross-training ensures that team members understand how their work impacts downstream operations. Certain "minor" variances in quality can have dramatic impacts later in the process. Second, it promotes flexibility. One employee can step in when another is on a break or can help reduce bottlenecks when a co-worker gets swamped. Third, as with the Northwest Natural Gas example, small teams that perform an entire task also lead to reduced batch sizes, reducing cycle time and making just-in-time methods possible.

THE ACCIDENTAL TEAM

Many organizations create de facto self-directed teams, intentionally or unintentionally, by reducing layers of management. For instance, Monsanto's Pensacola plant offered an early-out package in 1985 and lost most of its foremen. Since it was a down year, a decision was made not to rehire, but, rather, to implement self-directed teams. The decision paid off. By 1987, the plant produced the most A-grade product ever with the lowest staffing levels. In 1988, the plant broke its own record again.

By 1989 however, performance had leveled off and an assessment revealed that no one had defined the responsibilities of teams. In conducting a plant walk-through, we heard such comments as, "Things really haven't changed that much; the supervisor still makes most of the decisions."

This example highlights a common mistake. Organizations can show quick financial results by delayering, because they instantly release themselves from paying a tier of salaries. However, most of the work that the layer performed doesn't go away. And in the absence of clearly defined roles, the managers above the removed tier tend to assume the additional responsibilities.

This tendency to hoard responsibilities within the ranks of management occurs for many reasons. The invisible military/management paradigm often prevents people from even considering delegating the tasks to employees; and employees who have been well indoctrinated into the check-your-brain-at-the-door culture may believe that it is "not my job." The division between hourly and salaried tasks and union constraints also may get in the way. Remaining managers also may try to preserve their status; if they take on these new responsibilities they may be viewed as indispensable, or so goes their thinking. Interestingly, the employees may be reluctant to take on more responsibilities for fear of pushing their manager into the unemployment lines.

INTEGRATING SELF-DIRECTION WITH TQM

Total quality management and empowerment should be implemented in concert. Recent research confirms that these efforts should not be handled as separate, uncoordinated efforts vying for resources. [7] However, implementing them in tandem raises numerous questions. What should the effort be called? What should be implemented first?

How should the relationship between the various skills, techniques, training events, and the like be communicated? Let's address each issue in order.

Creating a Focus

The first strategic choice an organization has is whether and what to label the effort. If employees are jaded from past "programs," an organization may want to avoid giving the effort a name and a big kickoff. One city manager in the state of Washington just keeps harping on the need to improve customer service. Hewlett-Packard refers to the "H-P way." The advantage of this approach is that the organization avoids the program-of-the-month trap, in that it easily can incorporate new techniques and methods. The disadvantage is that it provides little guidance and quickly can dissipate with a change in leadership.

If an organization decides to pick a label to frame the effort, it is faced with many choices: customer service, total quality, empowerment, self-direction, high performance, just in time, and so on. Pick a label that will fit well with the culture and mission. For instance, "Quality is job one" was an appropriate rallying cry for Ford Motor Company, given the competitive challenges it faced from the quality conscious Japanese. However, a similar approach can fall flat in other settings. In health care, for instance, employees may be offended by calls for quality. Some may wonder, "What do you think we've been trying to do—kill people?" Instead, framing the effort as empowerment may yield a more positive reaction by valuing the contributions health care workers make to patient care.

Creating a Coherent Implementation Strategy

While there are an infinite number of ways to implement TQM and empowerment, there are three basic implementation strategies:

- Implement TQM and evolve toward self-direction.
- Implement self-direction and infuse quality principles and techniques.
- Redesign the entire workplace at one time.

The first strategy is probably the most common and is reflected in the organization of this book. A company begins by implementing TQM principles and training. Over time, the workforce improves its skills and knowledge. Empowerment begins as quality improvement teams (similar to quality circles), and it slowly evolves toward self-direction as the employees and their managers gain confidence. This approach ensures that employees are well educated about customer needs and quality issues before they are given significant power. The downside is that it may not fundamentally change the organization structure or practices.

Other organizations, especially in companies that frame the effort as empowerment, reorganize into self-directed teams. By putting the teams in touch with customers and providing appropriate quality focused team measures, the self-directed teams soon crave tools and techniques to help improve quality. The philosophies and techniques of TQM then can be introduced in response to their needs. This approach can improve employee morale and motivation quickly. The risk is that employees begin to view empowerment as the end, not the means to the end.

Some organizations prefer to redesign the entire workplace at once, usually using sociotechnical systems design methods as a foundation. [8] This approach involves analyzing both the technical needs of the work (e.g., process, physical layout, technology) and the social systems that impact the work and employee satisfaction (e.g., roles, rewards, empowerment). The workplace then is redesigned to optimize the technical and social systems. While this approach is used within existing organizations, we feel it is best applied when designing new plants or workplaces. The benefit of this approach is that it often yields quantum leaps in performance, since the entire workplace is redesigned at once. However, the intensive front-end analysis can leave organizations without the energy or resources to implement.

The approach selected will depend on many factors:

- *Competitive pressures.* Organizations faced with a quality crisis may choose to focus on quality.
- *Current levels of employee satisfaction.* Organizations with disgruntled employees may find empowerment attractive.
- *Management biases.* Managers who have faith in employees' abilities often will prefer empowerment; managers who question the value of involvement will prefer quality. This is also cultural. The Japanese tend to prefer a focus on quality, thus limiting empowerment. The Swedes, on the other hand, have embraced advanced versions of empowerment, including organizational democracy.

- *Organization size.* Small organizations usually offer employees broader job responsibilities, so empowerment may be less of an issue. Large bureaucratic organizations, on the other hand, may need to break what psychologists call "learned helplessness" before employees will fully participate in quality improvement efforts.
- *Internal knowledge.* While many organizations have people skilled in TQM, few have internal consultants with sociotechnical systems design experience. Familiarity often will drive the selection.

Communicating Interrelationships

Because American culture is event-oriented, we tend not to see interrelationships. This leads to situations like one witnessed in a chemical plant. The superintendent whined to the plant manager, "Last week we did diversity training; this week it's self-directed teams; next week it's statistical process control. I thought we were doing TQM!" If the superintendent was confused about the relationship between these efforts, you can bet other employees were, too.

It is helpful to create a simple visual that clarifies the relationships between the various techniques and buzzwords so all employees can see how the pieces fit together. Then make this visual model the centerpiece for all training, communication, measures, and the like. Figure 9–2 is an example.

OVERCOMING OBSTACLES TO SELF-DIRECTED TEAMS

Because self-directed teams represent a major change from traditional work settings, implementing them is not easy. In 1992, AXIS Performance Advisors surveyed over 200 people from a variety of industries to identify the biggest obstacles. [9] Five general areas emerged:

1. Mistrust of management's motives.
2. Lack of clear expectations.
3. Resistance to change.
4. Lack of participative skills.
5. Lack of executive commitment.

FIGURE 9–2
Relationships between Techniques—Sample Visual

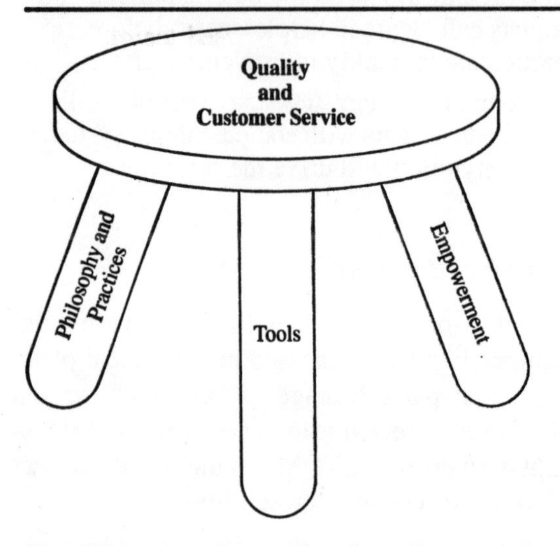

Note: This is a visual that the authors frequently use to demonstrate the relationship between related improvement efforts.

Source: AXIS Performance Advisors. Used with permission.

Mistrust of Management's Motives

Employees may mistrust management's motives for a variety of reasons. They may interpret self-direction as asking them to do more for nothing, or they may have heard that self-directed organizations require fewer people and they fear layoffs. In some organizations, management and employees have forged an adversarial relationship, often fostered by management capriciousness. If open communication has only been a pipe dream, overcoming this mistrust can take time.

So how do you build trust? Here are some suggestions.

- If you have the luxury of moving slowly, work on your managers' coaching skills before implementing self-direction. If managers are highly participative, trust will increase and the change to self-direction is much less traumatic.
- If you need to press forward quickly, two tactics will help. Create and communicate a clear, direct *statement of intent,* an explanation of why you are implementing self-directed teams. You should tie this to an open discussion of policies that relate to employees'

biggest worries (e.g., policies around layoffs, compensation, and so on).

If mistrust and skepticism are problems, take employees to other orga nizations that have successfully implemented self-directed teams. There they can talk to other employees and supervisors who thought this was just another program but now are convinced that it represents a better way to work. Employees are more likely to find credible those workers in other organizations, since they have no vested interest in obscuring the truth.

Lack of Clear Expectations

Lack of clear expectations may come from a variety of sources. No matter how many books people read, they remain vague on the details of how self-direction actually works. To make matters worse, there is no prototypical self-directed team because every organization must adapt the concept to its special needs. Some think self-directed means anarchy. Most often, however, the lack of clarity stems from not specifying the responsibilities that teams will assume. In many organizations, managers are reluctant to specify what powers they want the teams to assume, since the act of identifying the new responsibilities seems autocratic.

Unfortunately, since most employees have been indoctrinated into the Mushroom School of Management (keep them in the dark and feed them—how shall we say—compost), they view ambiguity not as an empowering opportunity but rather as unlimited risk. Employees fear that, if managers keep them in the dark about rules of the game, they can change them on a whim. Organizations that do not specify the team's authority may find managers hoarding the power they want or employees oppressing their managers. Instead, we recommend the following:

- Develop a detailed responsibilities list that specifies short- and long-range responsibilities the self-directed teams can assume. (See the example at the end of Chapter Ten.) Develop this list with representation from all levels of the organization. Let the teams select the responsibilities they want to assume first from the short-range options and coach them to ensure success.
- Develop a long-range plan that makes clear to all employees how long this process probably will take and what support will be provided along the way.
- Expose all employees to advanced examples of empowerment. The biggest limiting factor will be everyone's existing biases about what

is possible. Managers, in particular, tend to grossly underestimate what employees can do. Encourage employees to visit organizations that are further along than yours and share anecdotes, articles, and examples that will stretch their imaginations about how employees can participate in areas formerly the sanctuaries of management.

- Teach managers new competencies. Often the first-level managers undergo the greatest change and, understandably, they want to know exactly what is expected of them. However, no one can tell them exactly what their role will be. Since they are now in support of teams, not vice versa, their roles and tasks will change as the teams mature. All you can do is educate them on the competencies and skills that will make them successful (see Chapter Ten) and help them discover what their teams need from them.

  Few managers ever ask their employees, "How can I help you to be more productive?" You need to establish structured methods for this interchange to occur (such as upward appraisal surveys and "open appraisal" discussions—see Chapter Seven). Also, expose managers to peers in organizations so they can see varied examples of role redefinition.

Resistance to Change

People at all levels of the organization may resist for a number of reasons. They may not see a need to disrupt the status quo, or they may fear they will not be able to succeed in the new environment. And they may be unclear about what is expected. Some groups, especially of supervisors and technical professionals, will resist because their self-image has become intertwined with the power and special knowledge they have. Letting go can be a tough transition. Here are some useful tips:

- Develop a shared vision and values for the organization. Before employees get on board, they want answers to two questions: What's in it for me? and Do you mean me any harm? Vision and values help to answer those questions.
- Sell the problem more than the solution. Make sure all employees understand the business necessity for implementing self-directed teams. It's no accident that most organizations that have won the Malcolm Baldrige National Quality Award were motivated by crisis.
- Develop and communicate a statement of intent—why you are implementing self-directed teams. (See comments under Mistrust.)

- Separate layoffs from self-direction. If you must eliminate nonperforming managers or reduce staffing levels faster than the rate of attrition, do so first. Then bring the remaining people together to examine how self-directed teams can help them be successful with fewer people. Do not use self-direction as an excuse to deal with preexisting performance problems.
- Allow everyone to participate in redefining their roles. Help managers, union officials, and teams alike pick a preferred future and build their new roles around their interests as well as the needs of the organization.
- Help those employees who fear loss of status to see how, in this new team environment, their old roles and practices will be dysfunctional. For instance, in a traditional setting, maintenance employees gain significant status from repairing broken equipment. Their self-image as a kind of miracle-worker may inhibit them from sharing their knowledge with operators. They must come to see that fixing what is broken, while satisfying, is not as important as preventing breakdowns and sharing their knowledge.
- Plan to share with everyone the financial gains that self-directed teams will bring. Team-based compensation systems, such as recognition awards, gainsharing, and the like, can support the transition. [10] (See Chapter Eight.)

Lack of Participative Skills

Both managers and employees must possess participative skills for self-direction to succeed. While employees need to develop confidence and interpersonal team skills, managers' deficiencies typically represent a larger barrier. If managers are against self-direction and refuse to adopt participative behaviors, the effort has less than a 25 percent chance of success. [11]

Many managers think they are empowering when, in fact, they have abdicated their responsibility or are still maintaining tight control. The difference between how managers view themselves and how they are viewed by their employees is amazing. What we have here is a failure to communicate.

Few managers want to be autocratic. Many just can't imagine doing work any other way. They view certain tasks, such as budgeting, purchasing, and appraisals, as their responsibility and so they keep doing them. In fact, most managers think they *do* encourage participation, and few employees have the courage to tell them otherwise.

Managers must receive escalating feedback until they feel a need to change. Beginning with information about expectations and changes in roles, managers need increasingly pointed and personal feedback about their behavior. These approaches should help:

- Provide managers with participative skills training. It is critical that this training revolve around specific tough situations that they will encounter while moving toward self-direction. Understanding the theoretical concepts is only the first step; translating those concepts into appropriate actions is critical.
- Provide individual consulting for managers. Classroom training is insufficient. Observe managers as they interact with employees and provide private feedback.
- Implement official, structured ways for managers to get feedback. These may include upward appraisal surveys, feedback meetings, and the like. Unless trust is exceptionally high between managers, treat the data as private and confidential so it will be viewed by the recipients as developmental, not evaluative. Encourage managers to talk to their peers and employees about the data and to discuss their developmental needs with internal consultants or their own manager.
- After all reasonable efforts have been exhausted, remove or reassign any managers who are unable or unwilling to "get it."

Lack of Executive Commitment

Executives may lack commitment for many reasons. Since most executives came up through traditional ranks, self-direction can sound like some bizarre social experiment. They may not be familiar with self-directed teams and their effectiveness. They may have their focus elsewhere: fending off regulators, managing Wall Street, planning strategy. Oddly, many executives avoid dictating solutions, preferring to wait for good ideas to bubble up from the mid-managers. Many simply don't feel enough corporate pain to institute what they view as drastic measures.

Larry Miller, the author of *American Spirit,* writes about a culture curve. [12] He postulates that the culture of an organization or society begins to wane before its economic curve. The "blind spot" is the area where the culture has begun to fall before the financial health deteriorates. Here, executives are thinking, "Don't mess with success." They lay the foundation for the ultimate and precipitous decline of their organizations ifthey ignore the cultural side of their business. These executives think *their* organization doesn't need radical change, *their* performance will sustain them, *their* employees aren't ready to share responsibility for running the business.

Here are suggestions for gaining executive commitment:

- Prove the benefits. In an ideal world, a push for self-direction would come from the top. In the real world, the initiative often comes in an isolated pocket of the organization. If this is the case, insulate the self-directed team from outside influences until the team members experience success and gather data on performance improvements and their return on investment. Then communicate the bottom-line benefits to executives.
- Educate executives about the power of self-directed teams. In particular, make sure they understand that self-directed teams are not a new management fad. The research on which they are based (sociotechnical design) has been around since World War II. Since management as a discipline has only been around for about 100 years, that means the effectiveness of self-directed teams has been repeatedly demonstrated for almost half of that tenure. How many management fads can claim that?
- Remind managers and executives about the new competitive standards. America thrived after WWII because of productivity—our ability to produce the largest number of widgets for the lowest cost. However, now the competitive standards have changed. The new competitive standards—which self-directed teams can help achieve—are quality, variety, convenience, customization, and timeliness. [13] When people understand the game has changed, they often are willing to play by new rules.

CONCLUSION

Letting go is often the most difficult step toward improving quality and performance. Sharing power—responsibility, leadership, information, rewards—threatens the self-image and status of managers. However, without changing the power structure, gains will be limited. In this chapter we focused on self-directed teams as a way to share power. Their benefits are well-documented. All it takes is a manager's willingness to risk change and envision a more democratic, humane workplace where everyone has input into the matters that affect them most.

Tips, Tools, and Techniques
LESSONS TO LEARN FROM: RESULTS OF EMPOWERMENT

The following examples demonstrate the possible results from sociotechnical systems design and self-directed teams in a variety of work settings. [14]

Service and administrative

- Aetna Life and Casualty increased the ratio of managers to employees from 1:7 to 1:30 while improving customer service.
- Shenandoah Life processes 50 percent more applications and customer service requests with 10 percent fewer people.
- First National Bank of Chicago implemented teams and garnered steep increases in productivity, customer satisfaction, and staff morale.
- Corning Glass's computer center implemented self-directed teams and saved the organization $150,000 annually in salary-related costs while dramatically improving customer service and productivity.

Health care

- A medical center in Vancouver, Washington, reduced the cost of patient care while increasing customer satisfaction by implementing self-directed teams.
- Patient care teams were formed as self-managing groups in an Australian hospital and given responsibility for all direct care in areas having approximately 30 beds. Staff morale improved, statistics for average length of patient stay were reduced, and the teams gained in problem-solving ability.

Manufacturing

- Digital Equipment Corporation implemented self-directed teams years ago, resulting in a 40 percent reduction in process time, a 38 percent reduction in costs, and a 40 percent reduction in overhead. The teams also cut scrap to 3 percent.

- 3M turned around a failing division and tripled the number of new products with cross-functional teams.
- A chemical plant in the South allowed almost all of its foremen to retire one year and decided not to rehire but, rather, to implement self-directed teams. As a result, the plant turned around a quality related problem and produced more A-grade product than ever, resulting in 40–50 percent productivity gains.
- Procter & Gamble, which has experimented with sociotechnical systems for over 40 years, has found that its team plants are on average 40 percent better in all business indicators, including productivity.
- Zilog, Inc., used sociotechnical systems to design a new semiconductor plant in Idaho. The plant reached full production in 13 months instead of a more typical 25 months, circuit yields were improved by 100 percent, and employee turnover was reduced to only a few percent per year. Zilog subsequently used sociotechnical systems to improve white-collar productivity in the sales and engineering departments.

Public sector

- A school system in Boston improved test scores, reduced truancy, and increased teacher satisfaction by reorganizing around "houses" of 100 students. The teachers meet daily to discuss adjustments to schedules and student problems.
- A city in Ohio reorganized trash collectors into teams. They managed their own routes, improving customer service as well as the employees' quality of work life.

SUGGESTED READING

Carenevale, A. P. *America and the New Economy.* Alexandria, VA: ASTD, 1990.

Dumaine, Brian. "Who Needs a Boss?" *Fortune,* May 7, 1990, pp. 52–60.

Lawler, Edward. *High Involvement Management.* San Francisco: Jossey-Bass, 1986.

Lawler, E.; S. Mohrman; and G. Ledford. *Employee Involvement and Total Quality Management: Practices and Results in Fortune 1000 Companies.* San Francisco: Jossey-Bass, 1992.

Orsburn, Moran, and Zenger Musselwhite. *Self-Directed Work Teams: The New American Challenge.* Homewood, IL: Business One Irwin, 1990.

Chapter Ten

Management Beliefs

How to Align Philosophies and Practices

Total quality management is built on a foundation of beliefs and values. If managers do not internalize and act on these beliefs, TQM will fail. Inconsistencies in managers' words and actions lead to accusations of not "walking the talk." In this chapter, we will examine the new beliefs, roles, and competencies that managers must adopt to make TQM a success. We also will examine common problems that organizations encounter in getting managers to make this transition and how to overcome them.

WHAT IT MEANS TO WALK THE TALK

Top-management commitment often is heralded as a key factor in the success of any change effort. However, even with this "commitment," total quality management efforts can fail. This failure stems from management's unconscious assumption that this change does not affect the managers in any significant way. Few managers are adequately introspective and visionary to forecast the changes they themselves must make. These changes go far beyond behaviors and management "style"; they demand changes in thinking and assumptions that drive behavior.

Managers often do not perceive their own inconsistencies. For instance, when a fiberglass manufacturer was implementing self-directed teams to support its total quality effort, a vice president decided to sit in and audit the introductory workshops that were intended to educate employees on these teams.

During one workshop, an employee asked the vice president whether they, as team members, would ever be able to hire their co-workers. The vice president responded, "Probably not."

These two words said far more than the executive intended, and ironically, the employees were better able to decipher the real message than the executive himself. The two-word response was interpreted to mean that there were limits to this empowerment, that the executive did not truly believe in empowerment, that only the inconsequential responsibilities would be delegated. To the employees, "probably not" meant "no."

On the other hand, a manager who had internalized the new paradigm might have responded, "We don't know how far we can take this. What do you think? Do you think there would be value in hiring your own team members?" In the first response, control and dominance are paramount; in the second, mutual respect and continuous learning prevail. The beliefs and values they represent are light-years apart.

Real progress occurs only in tandem with management development. As Richard Tanner Pascale, author of *Managing on the Edge,* observed, the transformations of Honda, Ford, and General Electric were a direct result of their executives (Fujisawa, Peterson, and Welch) adopting new ways of thinking. [1]

Considering the number of actions that managers take each week and the number of words that emanate from their mouths, it is clear that managers cannot fake the new paradigm by knowing the "right words" or practicing the "right style." Slavish attention to techniques and behaviors will not yield satisfactory results. Managers must demonstrate integrity in management philosophy and practices.

However, consistency in direction is often more important than consistency in style, and that direction only comes from a coherent management philosophy. Paradoxically, sometimes managers must be autocratic about being participative. Kim Fisher, a well-known consultant in the field of self-directed work teams, describes two incidents at Tektronics, a high-tech Oregon company, which demonstrate this point:

> In the first, a manufacturing manager demanded very publicly in a precise, authoritarian manner that lower-level managers become less authoritarian. In the second, a manager absolutely dictated that nonmanagers be actively involved in the hiring of their managers. In each of the cases, the managers exhibited controlling, autocratic actions. While it is true that the autocracy was designed to create democracy, these behaviors in and of themselves were no less controlling than those of the stereotypical autocrat. And the people who were watching these incidents to learn how democratic managers acted—people who were observing the champions of high-commitment systems—were confused. [2]

Fisher also explains, "The plain truth is that the specific *actions* of good managers in these organizations often appear inconsistent with and sometimes even contradictory to what observers might consider 'correct' participative behaviors." [3]

OLD BELIEFS THAT GET IN THE WAY

In the last chapter we discussed the military/control paradigm that has become obsolete. There are many other beliefs embedded in traditional management practices that also must change. In this section, we will discuss several of the most common and damaging ones.

1. The Purpose of Business Is to Make a Profit

Since organizations in the private sector need to make a profit to survive, many executives have determined that making a profit is the purpose of their organizations. Numerous mission statements reinforce this belief. However, saying the purpose of business is to make a profit is like saying the purpose of your life is to breathe. Both are *preconditions* for continued survival. Neither represents the *purpose* for existence. Executives and managers shouldn't confuse the two.

This does not mean that financial performance is trivial. Far from it. However, our fixation on finance has led to many poor decisions. Manufacturing managers manipulate production to meet monthly efficiency targets, thus throwing their process out of control. Salespeople pressure customers so they can meet quotas. Research and maintenance expenses are postponed to make the annual report look good. Governments rush to spend leftover dollars before the fiscal period ends, whether or not that spending is wise.

Conversely, W. Edwards Deming believed that profits are the result of attention to quality and customer satisfaction, while the reverse is rarely true. Instead, he hypothesized the chain reaction shown in Figure 10–1. In a global economy, the purpose of business is not to make a profit. The purpose of business is to find profitable ways to meet and exceed customer needs.

2. Quality Costs Too Much

Conventional wisdom asserts that you get what you pay for; in other words, the higher the quality, the higher the cost. However, what customers are willing to *pay* is different from the *cost* of producing a product

FIGURE 10–1
The Deming Chain Reaction

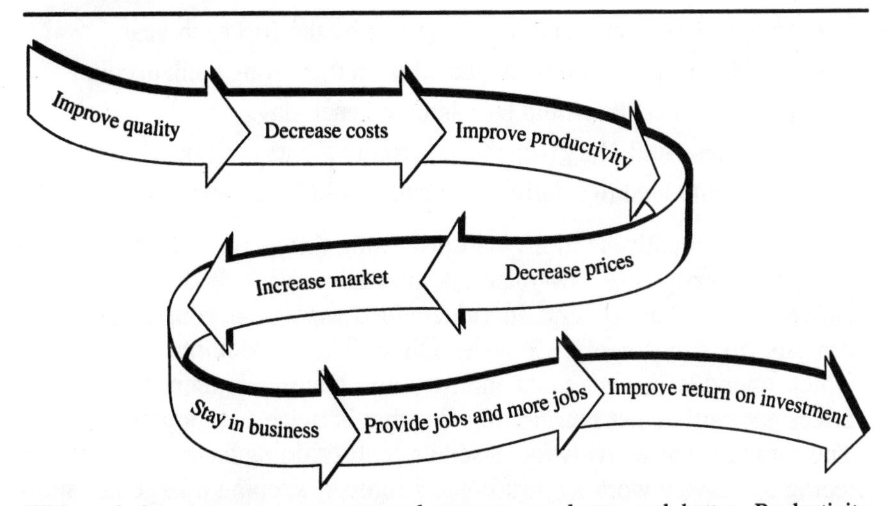

With each improvement, processes and systems run better and better. Productivity increases as waste goes down. Customers get better products, which ultimately increases market share and provides better return on investments.

Source: Peter R. Scholtes, *The Team Handbook: How to Use Teams to Improve Quality* (Madison, WI: Joiner Associates, 1988). Reprinted with permission from *The Team Handbook.* Copyright © 1988 Joiner Associates, Inc. All rights reserved. (Note that Joiner Associates added "improve return on investment," which Dr. Deming approved.)

or delivering a service. As Philip Crosby eloquently points out in *Quality Is Free,* pursuing quality should *reduce* costs. If an organization can eliminate most of the work associated with fixing what is broken or with reducing the hassles of getting work done, it can reduce time and costs significantly. Crosby estimates that the cost of this waste is 20 percent or more of sales in manufacturing companies and 35 percent of operating costs in service companies. [4]

Consider what your organization would be like if everything was done right the first time. You might eliminate the customer service department, which is typically a euphemism for customer complaint handlers. The division that repairs your products would be unnecessary. The time you spend walking purchase orders through the bureaucracy would be recovered. Imagine, you might even be able to reduce the price of your product or service and still deliver improved quality!

The naysayers have a comeback. Certainly you can't eliminate all waste or mistakes. Zero defects is an unreasonable goal. Those who are willing to accept less should consider these statistics if 99.9 percent quality standards were met:

- Two million documents would be lost by the IRS each year.
- 22,000 checks would be deducted from the wrong bank account.
- 1,314 phone calls would be misrouted each day.
- 12 babies would be given to the wrong parents each day.
- Two plane landings daily at O'Hare would be unsafe. [5]

Of course, quality is more than just doing things right. Due to market forces, higher-quality raw materials and employees tend to cost more. However, wise investments in value-added aspects of your organization can pay off handsomely. Consider Direct Tire of Memphis, Tennessee, which operates in a stagnant industry and charges higher than average prices yet earns a net margin on sales that is twice the industry average. The company has actively increased its costs of doing business by guaranteeing all service work against defects forever, keeping a large inventory of unusual tires, and renting a trailer to store snow tires in the summer.

> They don't grouse about high costs nor do they see spending as a way to differentiate—*they spend because it's the most profitable way to run the business.* It's profitable because the money they spend is on processes that transform everyone who enters the store into a repeat customer. Loyal, satisfied customers are glad to pay for the services they receive. . . . Repeat customers tend to spend on average about twice as much per visit as first-time customers who are responding to newspaper promotions (about $173 versus $90). But a first-time customer who has been referred by another customer spends an average $224 on the first visit! [6]

3. Employees Are Costs

Although most organizations now call their personnel department "human resources," thinking has not changed. Ask any manager, "What do you do when times get tough and the organization is losing money?" Most of the time, the answer will be to lay off employees. Companies that view their employees as assets, on the other hand, share the burden by reducing hours, sharing pay cuts, and the like.

The fact is that your competitors, in most cases, can buy the same technology, reverse-engineer the products, copy the services, and match the warrantees. What they cannot get without stealing your people is the learning behind those actions. Your employees must become true partners. They must have access to sensitive information and help set strategy. When educated on the strategic issues, they can become valuable resources.

The Johnsonville Foods story has been well covered but is worth repeating. In 1986, this sausage manufacturer in Sheboygan, Wisconsin, was asked by a competitor (for whom it had produced private-label products) to assume production for a plant the competitor was closing. While the offer was extremely attractive, the CEO, Ralph Stayer, was disinclined toward the offer, since it would require an enormous effort on the part of the employees. Instead of making the decision with his senior management, as he would have in the past, he took the issue to all the employees and asked them to consider three questions: What would it take to do this? What could be done to reduce the downside? Do we want to do it? Ten days later, teams returned with their response: Go for it! Since then, productivity in the factory has risen over 50 percent. Stayer describes his epiphany:

> I realized that I had been focused entirely on the financial side of the business—margins, market share, return on assets—and had seen people as dutiful tools to make the business grow. The business *had* grown—nicely—and that very success was my biggest obstacle to change. I had made all the decisions about purchasing, scheduling, quality, pricing, marketing, sales, hiring, and all the rest of it. Now the very things that had brought me success—my centralized control, my aggressive behavior, my authoritarian business practices—were creating the environment that made me so unhappy. I had been Johnsonville Sausage, assisted by some hired hands who, to my annoyance, lacked commitment. They had no stake in the company and no power to make decisions or control their own work. If I wanted to improve results, I had to increase their involvement in the business. [7]

4. You Work for Your Boss

In many organizations, the pyramid still has the point at the top. Yet since *Service America!* [8] was written in 1985, we all know the apex should be at the bottom, with managers supporting employees, not the other way around. But managers hire, fire, and appraise workers. The converse is rare. And few organizations have found ways to empower customers to provide direct feedback or rewards.

A top-down organization is built on an obsolete definition of leadership. To many, the role of a leader is to decide what needs to be done and then to rally support around the plan. However, most problems and situations today are too complex for any one person to have adequate knowledge or influence to solve them. As Larry Miller, author of *Barbarians to Bureaucrats* and *American Spirit,* so eloquently states: [9]

> The successful manager of the future will make full use of the collective wisdom of those within his jurisdiction and he will learn to derive pleasure, not from the making of decisions, but from assuring that the best possible decision is made.

Managers who internalize the inverted pyramid behave differently. In an automotive manufacturer, a "team coach" for a tool design department was successful in upending the pyramid. In a meeting with his less-enlightened peers, he admitted that he really didn't spend much time coaching his team anymore. Someone quipped, "Then we'd like to know what you do for eight hours a day." "Oh, the team always finds plenty for me to do," was his response. His peers were speechless. He truly worked for the team, offering to do whatever was necessary to help them be more productive. That is the essence of inverting the pyramid.

NEW ROLES FOR MANAGERS

So if managers shouldn't do what they used to, what *are* they supposed to do? In this section, we will discuss the new roles and competencies for managers and examine how to overcome the common obstacles to their development.

While high-performance organizations tend to be flatter, it is not always appropriate or prudent to furlough the individuals whose rung in the ladder has just been removed. In fact, these managers can be a tremendous asset in facilitating the change. After all, the teams need to learn what managers know, so coaching teams can become a full-time job for well over a year. Once relieved of many of their traditional responsibilities, managers can attend to more important issues for the organization's future.

There are four common roles that managers assume in a high-performance organization after the initial transition is complete.

Return to Front-Line Work

Many people are promoted into management for the wrong reasons: they had the longest tenure or were the best technician. Often people are attracted to the position because it is the only way to increase their pay and control over their work life. Some, when given the option of examining their true interests, return to the front lines.

Become a Technical Consultant

Those who were promoted because of their technical expertise often find a rewarding role in supporting the team with their technical expertise. These managers may stay in the chain of command, become part of the team, or be reassigned to a technical support area where the team becomes a customer.

Become a Sales Representative

Some teams become mini-enterprises—stand-alone business operations within a larger organization. They often have responsibility to make a profit through internal chargebacks or by selling their services to customers outside the organization. In these settings, the manager often becomes an account manager, generating new business, following up with customers, and the like.

Assume Higher-Level Duties

Most often, everyone in the organization assumes higher-level duties: employees do the work of supervisors, supervisors the work of managers, managers the work of executives, and so on. This frees executives to assume more valuable roles. Ask your executives how they spend their time and how they think they should be spending their time. The gap is usually gigantic. One vice president of a major utility responded, "We don't have time to do what I know we should—talking to customers, investigating new practices, chatting with employees. I spend most of my time in meetings—strategic planning, mostly."

THREE VILLAINS

Earlier, we said that managers can be an enormous asset to the transition. They also can be an incredible barrier. We have identified three common management personae to watch for.

Silent Saboteur

The most troublesome is the silent saboteur. This person nods and smiles and seems to go along, but, by omission and commission, he sabotages the effort. Instead of helping the team with a new responsibility, he may set the employees up for failure to "prove" they weren't capable. This person figures that the hubbub will pass by, going the way of all previous "programs," intending just to wait it out. The teams sense his lack of commitment and refuse to take risks.

Fence Sitter

This person isn't sure what to think and isn't sure which side to take. She wants to see which behaviors will be rewarded and, lacking the conceptual foundation, vacillates from overcontrolling and undercontrolling. As a result, the employees tend to withdraw, unwilling to risk initiative until they receive clear signals.

Oversupportive

This manager often is considered the paragon. She is incredibly helpful and supportive of the team and is intimately involved in everything it does. At least initially, the team loves her. Unfortunately, her ego is tied up in helping and being needed, which leads her to stifle the team's growth. Often the employees abuse this manager by selecting the most enjoyable responsibilities and delegating upward the unpleasant ones, such as dealing with nonperformers. Like children of an overprotective parent, employees with this type of manager never mature.

While some managers may fit cleanly into one category, most will display some of these dysfunctional behaviors at one time or another.

COMPETENCIES FOR EMPOWERING MANAGERS

Managers must learn that they now are primarily responsible for the workplace, not the work. While traditional managers focused on planning, controlling, staffing, and directing, the success of empowering managers will be dependent on another set of competencies, what we call the Four Cs®: coaching, consulting, cheerleading, and coordinating. [10] See Figure 10–2.

FIGURE 10–2
The Four Cs

Category	*Definition*	*Example Tasks*
Coaching	Helping others to improve their performance and make effective decisions.	Helping teams to make the best decisions through questioning. Observing performance and giving constructive feedback. Doing effective management by walking around.
Consulting	Giving information and training and diagnosing team dysfunction.	Investigating new technologies and bringing relevant information to the team. Intervening with troubled teams. Providing technical assistance. Explaining how systems work.
Cheerleading	Managing the climate, rewards, and culture.	Establishing meaningful team measures. Giving rewards and recognition. Helping teams celebrate their progress.
Coordinating	Aligning functions, eliminating barriers, negotiating with others outside your span of control.	Coordinating goals of various work groups and communicating across team boundaries. Improving alliances with customers and vendors. Recommending changes to business practices and systems.

Source: AXIS Performance Advisors, Inc. Used with permission.

Within these competencies, managers must learn new skills, including how to:

- Coach employees to think for themselves.
- Offer suggestions without them being interpreted as orders.

- Use management by walking around in order to build relationships and stay in touch without being perceived as a spy.
- Transfer complex knowledge.
- Redesign organizational systems to align with the values of teamwork and quality.
- Build strong and lasting alliances with others.
- Catalyze action without taking responsibility for the work.
- Build a shared vision.
- Manage interconnectedness within a self-directed environment.

The knowledge and techniques necessary to perform these skills are readily available. The challenge lies in getting managers to use these new skills. Four obstacles commonly prevent managers from pursuing their own development:

- Resistance—they don't want to change.
- Arrogance—they think they already are perfect.
- Skills—they don't have the skills to perform.
- Priority—they believe that other issues are more important to address.

GETTING OVER THE OBSTACLES

There are no magic wands to fix these four obstacles. However, we have a number of recommendations that may help.

Resistance

Resistance can take many forms: sweet smiles, passive-aggressive behavior, malicious obedience, and griping. Because demonstrating overt resistance is taboo in a traditional organization, it is easy to underestimate the magnitude of this problem. These tactics may help:

1. Establish an employment policy emphasizing that everyone's role will change but that no one will lose employment due to this change. At Corning, for instance, the company was able to rely on attrition to shrink the workforce while guaranteeing employment. [11] Make clear that support will be available to help people adapt, but that nonperformance will still be grounds for dismissal.

2. Sell the need more than the solution. Make sure everyone understands the pressures underlying the change.

3. Take managers on site visits to other organizations that are further along in the transformation process. There they will learn from disinterested peers that, while the change was difficult, the new role is far more rewarding.

4. Help everyone define a new, inspiring role for themselves. Ask them to imagine that half of their time has just been freed and to list what they would love to do with this time that would add value to the organization. Then ask them to pick the role that most excites them and encourage them to build a proposal around that new role. Remember that managers should be included in the redesign of their own jobs.

5. Provide training over time so they can be provided the skills they will need when they will most need them and have time to practice and get feedback on the job.

6. Implement "Autocrats Anonymous" meetings where managers can commiserate and learn from each other.

7. Use active, creative training techniques that draw them in, despite their resistance. Many consultants who offer workshops around self-directed teams and work redesign use simulations (like the star ship factory, which simulates a manufacturing process by making origami stars) to dramatically demonstrate the difference in traditional and high-performance organizations. Reference to those experiences often becomes part of the shorthand lexicon to describe the process. For instance, in one organization that had participated in a simulation involving the manufacture of puzzles, the rallying cry became, "Give the puzzles to the puzzlemakers!"

Arrogance

Perhaps it is just part of the human condition to assume that "I'm OK; everyone else is screwed up." Managers routinely underestimate the amount of changing they themselves must do. Organizations must set into motion escalating opportunities for managers to get feedback on their performance.

- Initially, the classroom setting will provide relatively nonthreatening opportunities to compare notes. Well-designed activities should reveal their strengths and weaknesses.

- In addition, however, on-the-job observation and coaching will be necessary.
- Formal upward appraisals (where the team assesses the performance of its manager) should be instituted. "Open appraisals" (see Chapter Seven) also can provide effective forums for continuous learning.

Skills

Few managers have all the skills necessary to be effective in a high-performance setting. The training they are provided must cover concepts as well as behaviors, but it should be addressed within the context of their real-world situations.

- Use vivid, relevant examples and cases so the managers can learn how to handle common situations as well as how to use the discrete skills. Use their real-world problems as a major source of content.
- Stage the delivery over time so they can get just-in-time training and have ongoing support.
- Provide numerous job aids to help them transfer the skills into the workplace. Since you are asking people to change habits, schedule specific tasks they must complete until the new practices become second nature. Worksheets, simple models, and other job aids also can be helpful crutches.
- Provide ongoing reinforcement of the techniques (e.g., discuss them in staff meetings and so on) and plan opportunities for them to get individualized coaching.

Priority

Often managers do not assimilate the new practices, because these are not perceived as adequately important. In the fray of everyday life, attending to these changes tends to come last. Here are some suggestions:

1. Get top management to model the behavior. When one plant manager was asked to demonstrate his commitment to completing the class assignments, he offered to send out a reminder memo. Instead, we suggested he ask his employees about their assignments, casually mention the results he was seeing from personally doing the assignments, display his Development Planner (a job aid) in his breast pocket, and discuss his own shortcomings and struggles.

2. Give managers explicit assignments to ensure the skills are transfered from the training room to the job. Encourage managers to complete the assignments at the beginning of the day. Managers often complain they do not have time to do the assignments; but if they perform them first, the old behaviors are the ones for which they will not have time.

3. Establish effective quality and team-oriented measures (see Chapter Six) and begin appraising managers on the progress their teams make.

4. Share success stories and promote only those managers with stellar participative skills.

CONCLUSION

Total quality management cannot be achieved without a drastic change in management philosophy. Managers must value employee involvement in the business, adopt new roles, and use new skills. In this chapter, we examined several beliefs that inhibit this transformation. The new roles and competencies for managers were explained as well as three troublesome personae. We also discussed ways to overcome the most common obstacles to getting managers to change.

Tips, Tools, and Techniques
DELEGATING THE RIGHT TASK TO THE RIGHT PERSON AT THE RIGHT TIME

The following Responsibilities List is a useful tool to determine appropriate tasks to delegate. It itemizes tasks that managers typically do. The list can be used to:

- Clarify existing levels of authority and empowerment.
- Negotiate future levels of authority and empowerment.
- Identify development needs of the team.

When managers examine the list, they often discover many tasks that their employees easily could perform. Other tasks often require some training or coaching before front-line employees can perform them adequately, but the benefits of involving employees in those tasks soon become apparent.

Responsibilities List				
Task	*Manage-ment*	*Team Leaders*	*Team Members*	*No One*
Setting goals				
Write business plan				
Set team goals				
Set individual goals				
Draft budget				
Approve purchase requests				
Select new equipment/tools				
Managing work				
Schedule work				
Manage priorities				
Promote safe work practices				
Monitor quality				
Approve time off/vacations				
Schedule time off/vacations				
Monitor attendance				
Staffing positions				
Assign work on daily basis				
Establish criteria for hiring				
Interview job candidates				
Hire new employees				
Orient new members				
Select team leader				
Decide on promotions				
Terminate employees				
Remove team members				
Conducting meetings				
Run daily start-up meeting				
Prepare agendas				
Lead problem-solving sessions				
Conduct safety meetings				
Conduct improvement meetings				
Record meeting results				
Coaching performance				
Establish expectations/standards				
Monitor performance				
Provide informal feedback				
Appraise team members				

Responsibilities List (concluded)				
Task	*Manage-ment*	*Team Leaders*	*Team Members*	*No One*
Appraise team leaders				
Appraise manager				
Provide on-the-job training				
Define training needs				
Schedule training				
Reinforce training				
Counsel troubled employees				
Provide progressive discipline				
Offer career guidance				
Communicate upward				
Rewarding results				
Give verbal praise				
Offer nonmonetary rewards				
Decide on monetary rewards				
Determine pay level of team				
Linking to others				
Propose ideas to management				
Interact with "customers"				
Interact with "suppliers"				
Coordinate with other depts.				
Communicate with management				
Other				

Follow these steps to analyze the Responsibilities List:

1. Add additional tasks to the list. Focus on tasks now done by managers that might be performed by employees. Include any tasks for which current authority is not clear.
2. Indicate who performs each task now. Checking the "Management" column means only managers perform the task. "Team Leader" means that

one person on the team, a specialist, performs the task. "Team Members" means that all employees can perform the task (including managers, if appropriate). "No One" means that no individuals are currently responsible for performing a task.

3. Identify who ultimately should do the tasks. This can involve simply going back through the list with a different color pen and checking the appropriate column. In situations where degrees of authority are critical, the following codes are helpful: A = Approve; R = Responsible; C = Consulted; I = Informed.
4. Once potential handoffs are identified (those responsibilities the team members do not do now but should/could in the future), break them into long- and short-term handoffs. Tasks requiring a significant amount of training or team maturity should be postponed until easier responsibilities have been successfully assumed.

SUGGESTED READING

Belasco, James, and Ralph Stayer. *Flight of the Buffalo: Soaring to Excellence, Learning to Let Employees Lead.* New York: Warner Books, 1993.

DePree, M. *Leadership Is an Art.* New York: Doubleday, 1989.

Fisher, K. Kim. "Managing in the High-Commitment Workplace." *Organizational Dynamics,* Winter 1989.

Geber, Beverly. "From Manager into Coach." *Training Magazine,* February 1992, pp. 25–31.

Johnson, H. Thomas. *Relevance Regained: From Top-Down Control to Bottom-Up Empowerment.* New York: Free Press, 1992.

Kinlaw, Dennis C. *Coaching for Commitment: Managerial Strategies for Obtaining Superior Performance.* San Diego, CA: University Associates, Inc., 1989.

Klein, Janice. "Why Supervisors Resist Employee Involvement." *Harvard Business Review,* September–October 1984, pp. 87–95.

Luke, Jeff. "Managing Interconnectedness: The Need for Catalytic Leadership." *Futures Research Quarterly,* Winter 1986, pp. 73(11).

Webber, Alan. "What's So New about the New Economy?" *Harvard Business Review,* January–February 1993, pp. 24–42.

Chapter Eleven

Structure

How to Design the Organization and Jobs to Promote Quality

Just as the architectural design of your home affects your behavior, organizational structure impacts employees. Organization design can inhibit communication, insulate employees from customers, and create significant barriers to innovation. Alternatively, organizations can be designed to improve quality, flexibility, and innovation. In this chapter, we will examine how organization and job design impact quality, and we will present principles to guide you in redesigning your workplace.

NEW STANDARDS FOR COMPETITIVENESS

IBM, once proclaimed a paragon, lauded in the book *In Search of Excellence,* has recently been brought to its knees. At the end of 1992 when the computer business rose 30 percent, IBM's empire continued to disintegrate. Forced to break its no-layoff policy, the hastily convened board of directors called for massive layoffs and restructuring for 1993. Shareholders revolted by calling for a company breakup similar to the breakup of Ma Bell. The stock price, having reached its pinnacle at a fraction below 176 in 1987, was barely above book value in the $45–50 range. Was the precipitous decline simply a function of poor management decisions or a reflection of a fundamentally changed climate in which practices effective in the past would now lead toward extinction? We believe the latter.

According to the Department of Labor, the driving competitive standard of the past was productivity: producing the largest volume of widgets at the lowest cost. However, those standards have changed with the globalization of our economy. The new competitive standards are quality, variety, convenience, customization, and timeliness. [1] These new standards all require flexibility. However, IBM's reluctance to cooperate with other organizations and its bureaucratic structure prevented it from responding quickly to market changes.

PROBLEMS WITH TRADITIONAL ORGANIZATION DESIGN

Restructuring is often management's first choice as an improvement strategy. Like a tile game, executives move boxes around on an organization chart, creating and eliminating divisions, centralizing and decentralizing, and changing the names of departments, most often with the intent to reduce costs. However, redesigning an organization while preserving obsolete paradigms is like rearranging the deck chairs on the *Titanic.*

Two main beliefs have driven traditional organization design. The first is the assumption that bigger is better: if you're not growing, you're dying. Success was defined as larger revenues, more employees, more business segments. Managers easily confused apparent economies of scale on paper with actual results. Communication lags and the hassles of doing business don't show up on financial reports, so organizations under pressure often centralize based on insufficient information. As Gifford Pinchot III, author of *Intrapreneuring,* explains:

> Seeing the waste, some call for more centralized controls, but the waste is not being created in inadequate controls. It is being created by removing the sense and fact of control from the only people close enough to the problem to do something about it. [2]

From a systems perspective, growth is not the goal; renewal is. Just as an animal will die if any part of its body (or all of its body) continues to grow far beyond it optimal size, organizations that pursue continued growth will self-destruct. Instead, survival depends on the ability to regenerate and to adapt to changing environments.

The second assumption is that birds of a feather should be put together, resulting in functional organizations. Accountants roost with accountants; engineers with engineers; customer service reps with customer service reps. This structure leads to stove-pipe organizations where issues must rise to the top before they can filter down to other departments. To get around this, some organizations decentralize; but even decentralized organizations often structure their divisions functionally, resulting in shorter yet equally impervious walls.

This preference for a functional structure has resulted in the rise of staff or "support" functions. Quality assurance and customer service became support departments, as if the functions they represented were not integral to manufacturing the product or delivering the service. As their power and specialization increase, these staff departments become more like guards and cops than like support functions. Furthermore, their members usually have a much greater affiliation with their profession than with the organization they purport to serve. The alternative is to design organizations around a whole.

NEW MODELS FOR ORGANIZATION DESIGN

The organization of the future will be designed by using different paradigms than those of the past. In an age where speed and flexibility are strategic assets, hierarchy must give way to flexible networks within and across organizations. Responsibilities must be delegated further down the chain of command, resulting in more decentralized structures. In *When Giants Learn to Dance,* Rosabeth Moss Kanter examines the "post-entrepreneurial" companies that are having success and states:

> In the post-entrepreneurial company, there are fewer and fewer people or departments that are purely "corporate" in nature; more responsibilities are delegated to business units, and more services are provided by outside suppliers. And fewer layers of management mean that the hierarchy itself is flatter. [3]

She identifies three main strategies that should guide the design of future organizations:

1. Restructuring to find synergies among pieces of the business, both old and newly acquired.
2. Opening boundaries to form strategic alliances with suppliers, customers, and venture partners.
3. Developing explicit programs of investment and coaching to stimulate and guide the creation of new ventures from within, what Kanter refers to as "newstreams."

Restructuring for Synergies

Restructuring often involves breaking bureaucracies into networks of autonomous business units. These smaller business units improve flexibility and empowerment. However, a network across unit boundaries must be maintained to gain synergy. The units must share learning and resources.

For instance, Eastman Kodak found its business languishing in the early 1980s and has worked to transform itself. Some of the key actions taken were to break the two divisions into four business groups whose officers were given responsibility for decision making and financial performance. This restructuring led to shorter lines of communication that made managers and employees alike feel they directly affected the business. Internally, the company encouraged "horizontal thinking," which helped integrate the segmented organization. This contributed to finding synergies across business units. These and other changes resulted in Kodak's return on equity rebounding to 19.0 percent by 1987, up from 7.5 percent in 1983.

This approach represents a significant change in thinking from what created conglomerates. As Kanter explains, true synergy is critical:

> The only real justification for a multibusiness corporation, in my view, is the achievement of synergy—that magical mix of business activities that are stronger and more profitable together than they would be separately. The "portfolio" or "holding company" approach . . . has been increasingly discred ited. [4]

Opening Boundaries

Organizations are discovering that they can increase capability without adding to their size by forming strategic alliances with other organizations. These may take many forms: joint ventures, spin-outs, partnerships, consortiums, and the like. While these relationships require significant energy to maintain, they provide organizations enormous flexibility.

For instance, under the direction of John Sculley, Apple Computer has relied on spin-outs. A spin-out differs from a spin-off in that it is intended to maintain a long-term relationship with the parent company, whereas organizations often spin off portions of businesses they no longer need. These spin-outs fostered empowerment and entrepreneurism, extending Apple's scope and strength without enlarging its girth.

Other organizations use closely aligned contractors. Benetton, an Italian clothier, owns few of the assets used to create its clothing. Instead, it contracts most of the work with small factories and retail licensees. The flexibility provided by this structure allows Benetton to promise fast turnaround for retail buyers, unheard of in the industry. Now buyers can purchase a small inventory of an item and see how it sells before ordering more. Not surprisingly, Benetton's customers are delighted with this capability because it helps them control their financial risk.

Stimulating Innovation

Larger organizations must find ways to stimulate innovation to compete. Product life cycles are waning in almost all industries, so the next product or service always must be in development. Organizations must encourage experimentation and cultivate promising prospects into newstreams—future products and services.

To develop newstreams, Kodak set aside 1 percent of revenues for new ventures and established "offices of innovation" at major facilities to encourage the entrepreneurial spirit within the organization. 3M, long known for innovation, encourages employees to devote 15 percent of their time to prove an innovation is workable, offering seed money and grants up to $50,000.

These three strategies—restructuring for synergies, opening boundaries, and stimulating innovation—form the basis for a far more flexible and customer-oriented business.

PRINCIPLES FOR DESIGNING A FLEXIBLE ORGANIZATION

Instead of oscillating between centralization and decentralization or growth and layoffs, organizations should establish an appropriate organizational strategy to meet their strategic needs. One strategy will not work for all organizations, of course. Organizations must establish a strategy that takes into account the nature of their business and industry. Then they must align the rest of the organization around that strategy. In particular, you must address eight dimensions of your organization, which we call the Eight Ss™. [5] These are shown in Figure 11–1 and discussed below.

Strategy. *Strategy* is the vision and plan of action that defines what an organization intends to achieve and how it intends to achieve it. Successful organizations have clear and elevating goals, which are communicated and instilled in the workforce while maintaining a customer focus and long-term focus.

Skills. *Skills* refer to the competence level of an organization's workforce. Successful organizations are devoted to continuous learning. They develop multiskilled employees by providing up-to-date technical and interpersonal skills.

FIGURE 11–1
Eight Ss

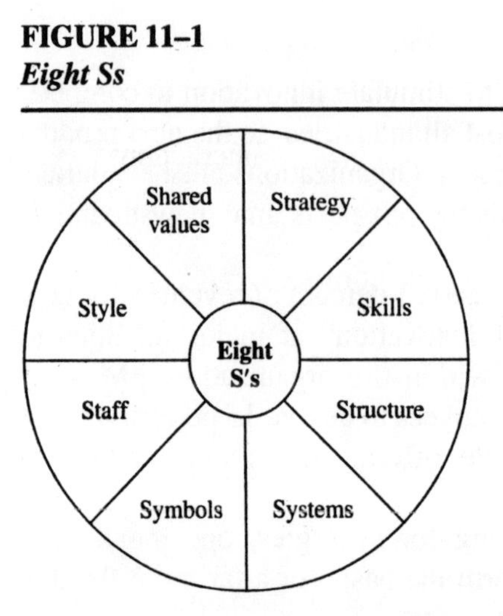

Source: Adapted from Richard Tanner Pascale, *Managing on the Edge: How the Smartest Companies Use Conflict to Stay Ahead.* New York: Simon & Schuster, 1990. Adapted with permission.

Structure. *Structure* describes the way a business is organized, how work is designed, and how responsibilities are assigned. Successful organizations maintain a flexible structure that promotes ownership and communication. Typically, they have few layers of management and limited support personnel. They often are organized around products and processes, instead of functions.

Systems. *Systems* include the mechanisms, work methods, policies, and procedures that support the operation of an organization. Beyond the work processes unique to the business, they include financial, information, human resource, and planning systems. Successful organizations have systems that are support quality and team work.

Symbols. *Symbols* are the overt or covert indicators of an organization's values and mission. Successful organizations have symbol consistent with their stated values. Within an empowering organization, these symbols must demonstrate egalitarian and democratic values and reinforce the continual pursuit of quality.

Staff. *Staff* refers to the characteristics of those people in key leadership positions within the organization. Successful organizations leverage diversity among their workforce and promote those who pursue the good of the entire organization, not just selfish interests.

Style. *Style* describes how managers behave in interactions with their staffs and their peers. Successful organizations have managers who are open, honest, communicative, and respectful. They value collaboration over competition, sharing power and managing conflict effectively.

Shared Values. *Shared values* refer to the common culture or values of the organization—how it views and deals with its resources, its people, and its community. Successful organizations set high standards of excellence, pursue continuous improvement, value diversity, and care about their impact on their communities.

These eight components must be aligned to meet the particular strategic challenges of an organization. The following five steps lead an organization through the necessary thought process.

1. Analyze Strategic Issues

Organization design should address strategic issues, so that first step in the process is to understand the strategic issues impacting the organization. Strategic issues tend to fall into five categories:

- *Macro.* Global and national issues affecting many industry segments.
- *Industry.* Issues unique to the industry segment.
- *Competitive.* Pressures that competitors place on the organization.
- *Customer.* Needs and demands that customers bring.
- *Internal.* Special issues within the organization.

Considered as a whole, these strategic issues imply directions for organizational design. At the macro level, for instance, the labor pool in the United States is becoming more diverse. To address this, an organization may chose to decentralize operations so that products and services can be targeted toward the various markets. Relating to industry issues, shortening product life cycles (as in the semiconductor industry) might imply a stronger need for spin-outs and newstreams. A new competitor with drastically lower labor costs may lead an organization to compete using high-involvement strategies. Internal issues, such as poor employee morale, might drive an organization toward greater participation. (See Strategic Issues at the end of this chapter for more information.)

2. Identify the Gap in Structure and Style

This strategic analysis should help you place your organization on the structure/style matrix (Figure 11–2), which defines how participative and how decentralized your organization must be to meet customer needs and remain competitive.

Organizations on the extreme end of the decentralization scale, using spinouts and other alliances, such as coalitions, joint ventures, and the like, often have these needs:

- To retain highly entrepreneurial talent.
- To control the risk associated with the venture.
- To share costs associated with the venture.
- To avoid regulations (as in the utility industry) that inhibit entrepreneurism.
- To enter new markets or industries.

Organizations on the extreme end of the participative scale usually have these needs:

- To react quickly to changes in the market or in customer preferences.
- To improve the quality of the product or service.
- To improve employee satisfaction and enrich jobs.
- To improve the quality of complex decisions based on a wide variety of input (as in concurrent engineering of automobiles or semiconductors).

Honda is an example of an organization at the upper right corner of the matrix, representing both high entrepreneurism and participation. It is considered one of the best-managed companies in the world. Takeo Fujisawa, the CEO, understood the importance of an appropriate, quality focused structure.

> Fujisawa made innovative use of organizational forms. Obsessed with the importance of maintaining Honda's leadership in automotive engineering and design, Fujisawa championed the idea of spinning off Honda R&D as a *separate,* wholly owned company with its own president. . . . Fujisawa's idea was a radical one; his colleagues found it so extreme that he encountered a great deal of internal opposition. . . . Later he spun off Honda Engineering (responsible for all of Honda's proprietary manufacturing machinery) as a second stand-alone

FIGURE 11–2
Structure/Style Matrix

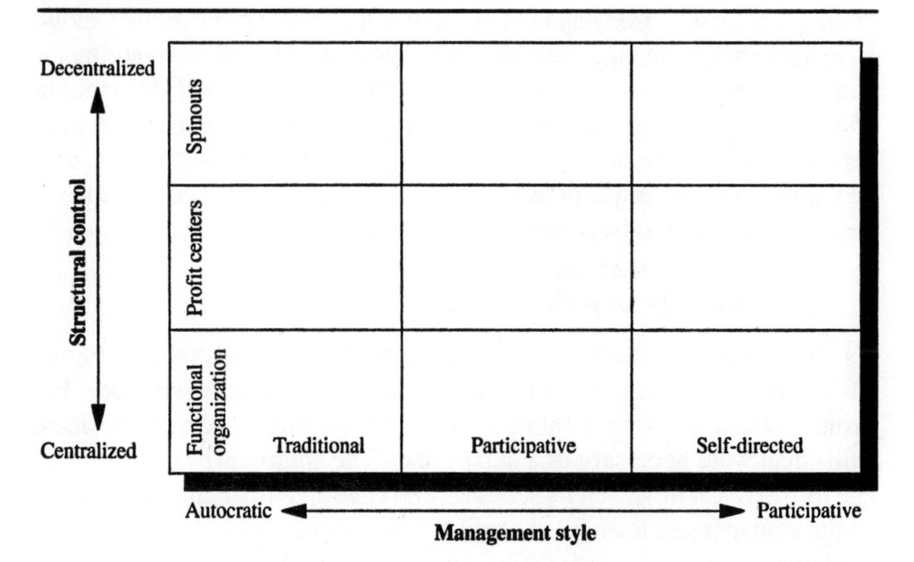

entity. This positioned Honda Manufacturing (with Sales, Marketing, and Administration functions) as the third leg of the stool. Predictably, these "three tribes," as they are called, generate a great deal of constructive tension. Whereas at General Motors, one elite and one point of view (Finance) holds sway, Honda is continually stimulated by these groups, each of which pushes and prods the organization from a different vantage point. To this day, all three maintain a distinct identity, have great pride in what they do, and are highly respected in their own right among competitors around the world. [6]

While Fujisawa's approach seems to violate what we were saying about functional organizations, Fujisawa overcame the drawbacks by creating a strong need to cooperate. The separate businesses were forced to view one another as customers.

Once you have placed your organization where you think you should be in the matrix, you can compare that position with where you think you are now. The distance between these two points defines how much of a transformation will be needed. All the remaining matrices should be viewed as overlays to the first diagram, so you can see the associated changes that will be necessary. These changes are additive; that is, as you move diagonally, you will need to satisfy requirements of both axes on the matrix.

When establishing organizational strategy, avoid two common mistakes. The first is uncontrolled expansion. So many organizations jumped on the "intra-preneurism" bandwagon in the mid 1980s only to discover they didn't have a clue how to manage new spinouts. The second mistake is ignoring the strategic importance of certain business units or functions. If divesting or underfunding a business unit is going to hurt your ability to pursue future markets, because the cost of learning and reentry would be prohibitive, you may be willing to accept lower or even negative returns. The core competence of the organization must be protected and magnified. [7]

3. Identify Appropriate Staffing and Skills

As you move from one box to another in the staffing/skills matrix (Figure 11–3), the individuals selected and the skills they possess will vary. For instance, those working within a spin-out will require more basic business skills than were necessary as a department. The spinout will have to manage its own cash flow, sell its services, and negotiate financing with banks, so the management team must acquire those skills.

However, not any perspective and background will do. The people selected to handle new functions, such as accounting and marketing, within the spinout will need to have a small business perspective, assets that many within the host organization may not have. One training department spun out by an electric utility failed to understand this distinction, staffing key positions with individuals accustomed to larger organizations. The new management team acquired company cars, fancy offices, and large entertainment accounts. Predictably, the business encountered significant cash flow problems and within 18 months drastic cutbacks in personnel and spending habits were forced.

At the self-directed end of the matrix, individuals will need team as well as technical skills. They will need to be competent in group process, group problem-solving techniques, consensus building, and the like. Promotions and hiring should be done on the basis of both sets of skills.

Again, Honda provides a useful model for understanding staffing and style needs. In a bold move, it recruited young and inexperienced employees for its research and development group, believing that their enthusiasm and lack of preconceived ideas would be assets. An executive described how Honda managed the risk:

> We control the *task* of researchers quite tightly, but loosen *controls*. Ideas with great potential often emerge. We monitor the creative breakthroughs carefully,

FIGURE 11–3
Staffing/Skills Matrix

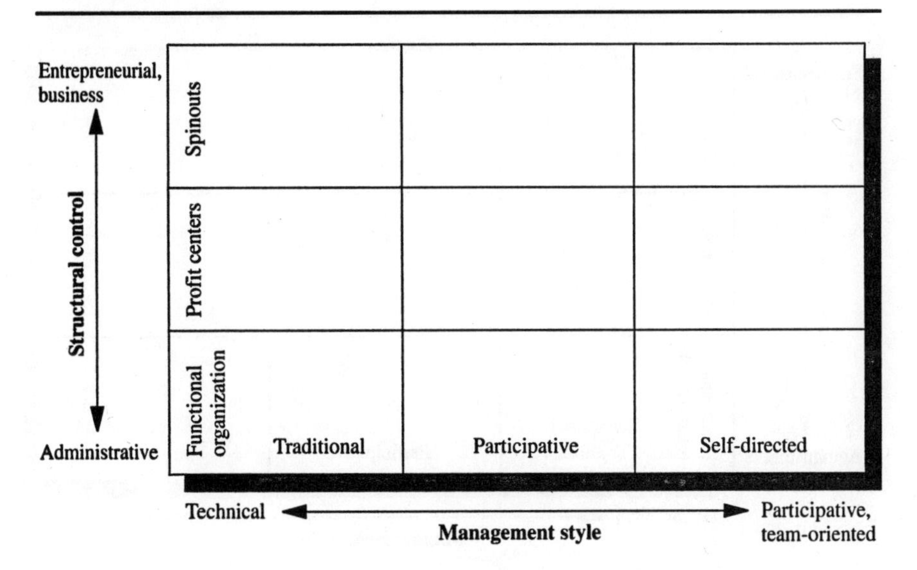

and when they look promising we move fast to develop them. Too much freedom doesn't work. But we take the chance of giving young researchers basic goals and responsibilities, and then letting them go by themselves. [8]

To match the staffing strategy, Fujisawa promoted a team-oriented style. Managers were encouraged to be coaches by staying involved and focusing on asking good questions instead of directing. Also, to overcome cultural conditioning, rank was disregarded in the conflict resolution process.

4. Align Shared Values and Symbols

In the shared values/symbols matrix (Figure 11–4), the values and organizational symbols should be aligned to support the organization's strategy. Spinouts will need to foster risk taking, innovation, and a customer focus. These values can be supported through such symbols as stock ownership, skunkwork settings, customer satisfaction measures, and funds for pet projects.

At the self-directed extreme, symbols and values must support teamwork and participation. Perks should be assigned based on need, not hier-

FIGURE 11–4
Shared Values/Symbols Matrix

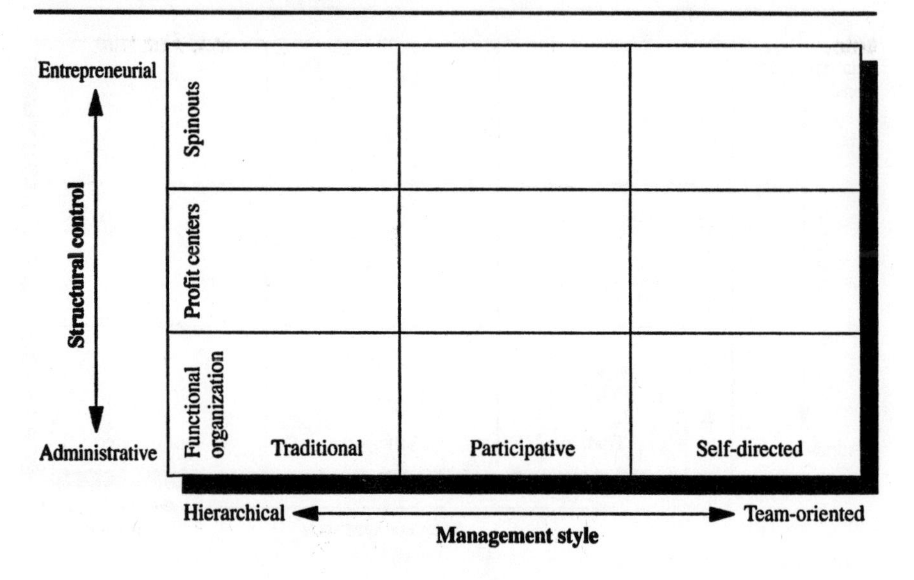

archy, which may lead to the secretary having a larger office than the manager and the delivery clerk having the assigned parking spot instead of the plant manager.

For example, Honda is strongly guided by its golden triad: enduring values, trust, and empowerment. To ensure that senior executives stay in close touch with each other, their employees, and customers, the top 40 directors are housed in an open area with only six desks, five conference tables, and 11 secretaries. [9] This symbolic setup—an executive version of musical chairs—forces them to interact and to stay in touch with the real work of the organization.

5. Align Organizational Systems

The last step in the process is to align the organizational systems with the chosen strategy and values. Organizations moving toward greater decentralization may need to allow for greater flexibility in their human resource systems so the organizations can operate under different compensation and benefits structures. Integrating planning across these more-or-less independent businesses will become a major challenge if the organization is to gain benefits of synergy. These alliances require significant attention by executives to keep them aligned.

For more participative organizations, such human resource systems as selection, appraisal, and compensation must be redesigned. For instance, when Boeing opened a high-involvement plant in Spokane, Washington, its selection procedure involved four weeks of training *before* a candidate could be hired, and at least one-quarter of that time involved developing the applicant's team skills. (See Chapters Seven and Eight for appraisals and compensation, respectively.)

Other systems also will need to be changed. Planning may need to involve new people. Information and reports may need to be routed differently. Procedures for purchasing and budgeting may need alteration. (See Chapter Twelve for more information.)

PROBLEMS WITH TRADITIONAL JOB DESIGN

The design of jobs within the organizational structure can be just as important to achieving quality as the macro organization structure. Within the realm of job design, rigid job classifications and corresponding job evaluation systems box people into positions that restrict what individuals can contribute to the organization. Often unions have fostered narrow job classifications and filed grievances about crossing lines of demarcation. Ridiculous rivalries sprouted, leading to absurd cases where one worker would not hand another a wrench or a screwdriver because he was not a pipefitter or an electrician. Within government, the civil service system is accused of similar effects. Job descriptions, the most visible artifact, and the job evaluation systems they represent require enormous energy to maintain while they reduce flexibility.

Management also limited the contribution of employees. Since the turn of the century, organizations have operated on Frederick Taylor's model of separating the thinking about work from the doing. Managers were responsible for making decisions and solving problems. Employees were expected to be good drones, toiling thanklessly for the queen bee.

Work redesign and "productivity improvement" efforts often were done *to* employees, not with them, as if employees were lifeless cardboard cutouts. But as Rudy Perpich, governor of Minnesota, noted:

> Efforts to improve productivity usually undermine both productivity and morale; efforts to improve morale by empowering employees usually heighten both morale and productivity. [10]

While many organizations may not want to follow in the footsteps of some organizations where only one job title is used for all employees, we certainly need new principles for designing jobs that will leverage all the assets people bring to the workplace.

DESIGNING QUALITY JOBS

Just as poor organization design can limit the performance of groups, poor job design can limit individuals. These five principles that follow should guide team and job design.

Organize to Maximize Customer Satisfaction and Quality

Work can be organized around many principles: by function, product, process, customers or market segments, project, geography, and so on. Organizations often focus on internal efficiency when choosing how to organize. However, this rarely results in the best choice for quality and customer satisfaction. For a time, IBM was organized around product groups, so a customer needed to call a different sales representative for a Selectric typewriter, a minicomputer, or a mainframe. Internally, this decision made sense; after all, who could be knowledgeablc on all of IBM's products? But customers complained that they wanted one rep who could help them figure out what they needed, rather than have to figure it out for themselves so they would know who to call.

Similarly, most governmental agencies are organized around programs, not missions or customers. In *Reinventing Government,* the authors describe an at-risk youth (pregnant, on welfare, and with a juvenile record) who had more than half a dozen caseworkers in various agencies.

> The system was so fragmented that, while each agency was performing a discrete service, no one was dealing with her real emotional needs. [11]

Pick an organizational strategy that will maximize employee ownership with the quality of the product or service and establish obvious links to customers. For instance, at Monsanto's Pensacola plant, the company broke the large plant into three product groups and those product groups into independent product teams. The teams view others within the plant as internal customers. The service product group represents those functions for which each team did not need a full-time person. Operational planning is done in two rounds: first by the line teams, and then by the service teams (which represent staff functions) based on the explicit needs of the line teams.

Put Interdependent People Together

Make sure that those who must work together on a regular basis are on the same team and working toward the same goals. This may lead to dramatic changes in organizational structure and role definition for specialists like engineers, quality assurance technicians, and maintenance workers.

For example, at Monsanto's Pensacola plant, accountants have been assigned to teams and play an active role in team meetings. They now view their role as helping the team manage its costs, rather than providing timely and accurate reports. In their on-site power plant, for instance, the accountant helped team members assume responsibility for managing cost centers worth millions of dollars. The awareness that the line workers have gained has led to dramatic cost-saving measures, and the accountant now has a tangible understanding of where the numbers come from.

If the team does not require a full-time member, the supporting staff organization should establish customer-supplier relationships with the line teams. A manufacturer in Phoenix, for example, has assigned purchasing people within the purchasing department to certain line teams, placing them physically close to the teams. This customer-supplier relationship and physical proximity have led to improved purchasing decisions and faster turnaround.

Provide Meaningful Feedback and Consequences

One of the advantages of organizing teams around a whole piece of work is that it facilitates performance feedback. At Volvo, for example, a team that assembles a whole car can get information about the performance of *its* cars. To the degree a work group completes a whole product, process, or project, or serves a customer, the members receive better information about the quality of their work.

People at all levels of the organization need better feedback about performance. To illustrate, Xerox requires its executives to monitor the customer hot line. This gives the executives real-time feedback on the pulse of their customers.

In some cases, the customer may even participate in the manufacturing of a product or the delivering of a service. For instance, one manufacturer of CIM (computer integrated manufacturing) equipment puts a customer on the team. This gives the team the flexibility to change the specifications of the system up to the day before shipping.

Feedback should be provided to vendors as well. At a food processing plant, for instance, teams use a card system to communicate quality problems to their vendors. The teams expect the vendors to respond to the quality problem within a specified time. If a vendor fails to live up to a team's expectations, the team has the power to change vendors.

Eliminate All "Bad Jobs"

Every organization has its "bad jobs," the tasks no one wants to do. They may be messy, dirty, stressful, tedious, or demeaning. In the past, these unpleasant tasks have been aggregated into single positions, and then we went in search of people desperate enough to fill them. Instead, you should seek to redesign the process to eliminate these tasks (because they often represent waste) or, barring that, to share the responsibilities.

For instance, in the human resources department in Emanuel Hospital, the receptionist position was the bad job. The human resources staff reorganized around customer groups and rotated responsibility for the reception area and phone coverage. The receptionist then assumed many of the more challenging human resources functions, dramatically increasing the capability of the department. [12]

Automation and technology often can eliminate bad jobs. The advent of electronic mail, for instance, has eliminated the need for hoards of receptionists. However, this trend has a dark side. As Barbara Garson documented in *The Electronic Sweatshop,* many professional positions are being deskilled and clericalized. Expert systems now tell money managers what to buy for a portfolio, fast food managers what to make, and welfare caseworkers what actions to take. Professional work in many organizations is being replaced by mindless data entry. In Sweden, by contrast, the welfare workers refused to use computers unless all the time saved was used for true social work.

If a bad job cannot be eliminated, it should be shared. To illustrate, a team may share "housekeeping" tasks or rotate the responsibility for taking minutes of the meeting. Operators may be expected to clean up and maintain their own equipment so no one has to clean up pools of oil or other messes they didn't create.

Team Members Should Be Multiskilled and Manage Themselves

In a traditional organization, employees work within narrowly defined job classifications. This enables managers to quickly train and replace workers—standardized parts in a human machine. However, the demographics of our society have eliminated the steady stream of available workers. We must make the most of the employees we have. Skills can be increased horizontally (i.e., learning what other co-workers do) as well as vertically (e.g., learning management tasks). In most cases, the flexibility and performance improvements that come from this cross-training more than offset the costs.

One note of caution: Don't force cross-training where it doesn't fit. Cross-training does not necessarily mean that everyone should do all jobs. In many cases, some positions (e.g., nurses, engineers, brokers) require so much additional training or experience that cross-training in all functions does not make sense. However, often those holding the specialist positions can relinquish or share many of their responsibilities and should be encouraged to do so.

Also, some positions are so labor-intensive that self-management is not practical. These are the bad jobs, which still use people as if they were machines. In time, they will evolve into positions that use the capabilities humans possess. For example, one manufacturer had a position where the employees were practically chained to their machines for their shift, leaving no time for team meetings or problem solving. In these cases, some free time must be found to allow self-direction. Solutions include adding people, using temporary help, paying overtime, or redesigning the process and technology to eliminate the need for constant human vigilance. Before this free time is created, pushing people toward empowerment will only cause frustration.

TITEFLEX—AN EXAMPLE OF JOB REDESIGN

Best-selling author Thomas Johnson, author of *Relevance Regained,* has described Titeflex, an industrial hose maker, which provides an excellent illustration of these five job design principles. In 1988, they were implementing an MRP (materials requirements planning) system to handle the paperwork associated with manufacturing. Employees were disgusted with the number of meetings necessary to get work done, and order entry took between three to five weeks. Once an order was entered, it passed among 12 departments, including a quality assurance department staffed with 50 people. This maze added another six weeks to cycle time.

When Titeflex redesigned the work, they established the "Genesis Team," which consists of six people from all relevant functions to process new orders. One team member is responsible for staying in touch with the customer and acting as liaison with the factory cells. Their results are impressive. Order processing now takes between 10 minutes and a few days. Manufacturing cycle time ranges between two days and one week. [13]

All five job design principles covered in this chapter were applied at Titeflex. The Genesis Team sat together, included all interdependent parties, and was linked directly to the customer from whom they could get meaningful feedback. Such bad jobs as order entry to feed the MRP system and numbing meetings were eliminated. Team members became cross-trained in quality and scheduling tasks. Implementing these principles resulted in a reduction of span time from 11 weeks down to 1, providing compelling evidence of the impact of job design on competitiveness and quality.

CONCLUSION

Organization and job design significantly impact an organization's ability to respond to competitive challenges. In this chapter, We presented three strategies for designing flexible, adaptive organizations: restructure for synergies, open boundaries, and stimulate innovation. We presented a model for deciding how decentralized and participative your organization should be. We also presented five principles for designing jobs to promote quality and empowerment.

Tips, Tools, and Techniques
ALIGNING STRATEGY AND ORGANIZATION DESIGN

Strategic issues should drive your organization design. To place your organization on the structure/style matrix, you must weigh strategic issues that fall into several broad categories. Under each category, we have listed issues that would imply a need for a more decentralized structure and a more participative style.

Macro Issues

Factors that imply a need for more decentralized structures include:

- Globalization and demassification of markets.
- Uncompetitive labor costs.

Factors that imply a need for more participative styles:

- Baby boomers demand more control over their work.
- A better-educated workforce.

Industry Issues

Factors that imply a need for more decentralized structures include:

- Regulations (as in the utility industry) that inhibit innovation.
- Mature industries in need of "newstreams."
- A need for a disinterested party (e.g., a consortium or trade association) to gather benchmark data.

Factors that imply a need for more participative styles:

- Competition for talent in the applicant pool.
- Processes requiring complex decision making in which no one person or set of people has enough information to make decisions (as in concurrent engineering of products).

Competitive Issues

Factors that imply a need for more decentralized structures include:

- Competitors who are able to beat your quality.
- A need to pool resources among some competitors to share the risk or cost of new ventures, research, and so on.

Factors that imply a need for more participative styles:

- Competitors who are able to surpass your flexibility.
- A need to react quickly to changes in customer preferences.

Customer Issues

Factors that imply a need for more decentralized structures include:

- Demand for faster and more flexible delivery systems.

- Demand for more product or service variety that is customized to certain markets.

Factors that imply a need for more participative styles:

- Need for the product or service to be adapted to the customer's needs at the point of purchase.
- Highly variable customer needs (as in many service businesses).

Internal Issues

Factors that imply a need for more decentralized structures include:

- A need to retain highly entrepreneurial talent.
- Poor relationships between interdependent groups.
- Business operations that serve vastly different markets, pull employees from different pools, or use different processes.

Factors that imply a need for more participative styles:

- Low job satisfaction.
- Bad jobs.

SUGGESTED READING

Hammer, Michael. "Reengineering Work: Don't Automate, Obliterate." *Harvard Business Review,* July/August 1990, pp. 104–12.

Hanna, David. *Designing Organizations for High Performance.* Reading, MA: Addison-Wesley Publishing, 1988.

Kilmann, Ralph, and Ines Kilmann. *Making Organizations Competitive: Enhancing Networks and Relationships across Traditional Boundaries.* San Francisco: Jossey-Bass, 1991.

Kolodny, Harvey, and Barbara Dresner. "Linking Arrangements and New Work Designs." *Organizational Dynamics,* Winter 1986, pp. 33–51.

Nadler, David; Marc Gerstein; and Robert Shaw. *Organizational Architecture: Designs for Changing Organizations.* San Francisco: Jossey-Bass, 1992.

Pasmore, William. *Designing Effective Organizations: The Sociotechnical Perspective.* New York: John Wiley & Sons, 1988.

Chapter Twelve

Systems

How to Leverage Financial, Information, and Planning Systems

Organizational systems are powerful predictors of human performance. They are often the last to change, however, creating powerful incentives to maintain the status quo. For TQM to succeed, an organization must redesign these systems to support quality. Systems built on antiquated management philosophies must be eliminated and others, which have only helped to manage bureaucracy, must be changed to add value to the customer. Since earlier chapters have addressed key human resource systems, this chapter will address how to modify financial, information, and planning systems to improve quality and empowerment. We also will present a process for redesigning any business system and show how this process was used to change an auto manufacturer's business planning system.

HOW ORGANIZATIONAL SYSTEMS INFLUENCE PERFORMANCE

Organizational systems are artifacts of management philosophy. They embody the values and beliefs of their architects. Perhaps because they are artifacts, they are followed with a certain reverence, and organizations seem reluctant to change them. In a turbulent world, this leads to troublesome rifts between what an organization needs from its employees and what it gets. Take budgeting, for instance. In many organizations, this system is used both to plan and to control. The assumptions behind budgeting include:

- Managers will ask for more than they need.
- Managers will abuse the use of funds if they are allowed to shift money from one line item to another.
- Next year should be a natural extension of last year.

Consequently, the typical budget system uses across-the-board cuts to reduce costs, reports variances by line item to control abuse, and takes funds away if they were not used in the previous year. Managers quickly learn that the politics associated with fighting for dollars outweighs the mechanics of budgeting.

This leads to perverse behavior. Equipment is rented for more than its purchase price, because the capital budget was depleted. To meet head-counts, employees are laid off and rehired as consultants at three times the pay. Materials are purchased, because they were budgeted even though more cost-effective solutions existed. Near year-end, managers spend wildly to justify similar funding levels for the following year. Government organizations are often the worst offenders.

The city of Fairfield in northern California decided this was a crazy way to operate within the antitax Proposition 13 environment. The city implemented a general fund approach in which line items were used only by the managers themselves to track expenditures; the council never saw or voted on them. Funding was based on a formula that included previous funding levels, inflation, and population growth. The managers were free to move money between line items as they deemed appropriate.

The flexibility to move and save money changed behavior. When the city investigated the cost of building a weather covering over a gas pump, the architect gave a price tag of $30,000. Thinking that an absurd cost, the city purchased glass-enclosed bus stop covers for $2,500 instead. Since making this change, Fairfield has been named one of California's most fiscally sound cities, spending $6.1 million less than it appropriated in 1991.

The necessary changes to systems often seem radical. A director within Fairfield described his reaction to the new budgeting system in this way:

> "When I came here from Sacramento, my instinctive reaction was that this system was crazy," said finance director Bob Leland, who previously worked in the state finance office. "It was totally alien to what I had experienced. But I got converted in a hurry. I wouldn't trade it now." [1]

Similar overhauls will be necessary with all our organizational systems to maximize empowerment, flexibility, and quality.

FINANCIAL SYSTEMS

Prior to the 1950s, financial systems generally were not used to drive the organization but rather to track the results. Instead of defining success, they monitored it. With the popularity of divisional structures, strategic business units, and conglomerates, however, financial reports became the basis of making business decisions in many organizations. Corporate executives became more like money managers than leaders of the operation, more concerned with return on equity than process improvements and markets. Thomas Johnson, author of *Relevance Regained,* likens this to playing a game while keeping your eye on the scoreboard.

In Chapter Six we discussed how financial measures should be supplemented with customer and quality data in the boardroom. But overreliance on financial measures is only one of the problems. Existing accounting practices often lead managers to make poor business decisions.

Unit Cost

Traditional accounting methods have spread fixed costs evenly across units of production or sales. In a manufacturing setting, this encourages plants to increase production, regardless of whether there is a market for the increasing inventory in their warehouse. As plant managers struggle to meet the numbers mandated by their executives, they become increasingly nervous near the end of the financial period. In an attempt to meet targets, they speed up production, postpone maintenance, and work employees overtime. This leads to at least three problems. First, unless there is unlimited demand for the product, the organization can end up with excess inventory. By artificially speeding up the manufacturing process (instead of redesigning it to reduce cycle time), the process is thrown out of control; and before a process can be improved, it must be in control. Third, employees become disgruntled, having been ordered to work harder and faster, not smarter.

Instead, organizations should work toward a just-in-time (JIT) system, which tries to manufacture what the customer wants when he or she wants it. This can be achieved only by focusing on reducing cycle time (the time it takes to make or provide what the customer wants), not focusing on unit cost.

Cost Allocation

How costs are allocated can lead to major distortions of information. For example, assume an organization produces the same amount of Product A and Product B. However, Product A requires twice the number of purchase orders to manufacture, demands twice the advertising expense, and generates twice the volume of telephone calls on the customer hotline. Standard accounting practices would apply the same amount of overhead to both products, while it is clear that Product A actually is absorbing far more organizational resources. In this example, standard accounting methods may lead managers to believe that Product A is profitable when, in fact, it may not be. Similar distortions occur when volumes vary dramatically.

Activity-based costing (ABC) and its cousin, activity-based management, can account for these differences, helping organizations get a more accurate view of the real cost of a product, process, or customer. It can show the entire cost associated with a decision (e.g., an engineering change), not just the impact on the department in which that decision is made.

However, ABC accounting methods represent only a partial solution. This information must be viewed through the lens of quality. Johnson describes two organizations that used ABC to make poor decisions. An auto components manufacturer discovered that small lot sizes cost more than large ones, and so it cut prices on large lots and discouraged sales representatives from taking orders from customers who wanted small lot sizes or frequent deliveries of small quantities. The other organization, a manufacturer of personal computers and electronic testing equipment, discovered that auto-insertion machines cost less than manual insertion techniques, even though it required that components be spaced further apart. The engineers opted for the auto-insertion machines.

> In both of these cases, the ABC information prompted managers to reduce costs and improve short-term profits by altering product mix or process mix, not by altering the way work is performed or customers are served. . . . But in the long term, both companies made choices that were likely to impair their competitiveness and profitability.
>
> To achieve competitive and profitable operations in a customer-driven global economy, companies must give customers what they want, not *persuade* them to purchase what the company now produces at lowest cost. [2]

Relevance and Usefulness

As organizations become more quality focused and team oriented, they demand different information from accounting and finance departments. When a division of General Electric implemented just-in-time, their accounting systems were no help:

> Ironically, when plants in GE's Medical Systems Group first moved toward JIT in the mid-1980s they discovered the corporate accounting systems could not produce information about the total actual costs. Now these plants bypass the

> accounting system and track total payroll . . . and total material costs at the plant level. [3]

Similarly, accounting methods rarely can provide budget data on a team level. As with the GE example, employees are forced to have redundant systems to get the information they need, how they need it, when they need it.

TQM organizations ask different questions of their numbers. Instead of market share, they may want to know "customer share": how much of a product the customer buys from their organization. They many want to know costs broken down by customer, not units. They may need profit and loss reports by team. Most existing accounting and financial packages are inadequate to meet these demands.

Timeliness

Timeliness is another vexing problem within most accounting and financial systems. Closing the books monthly is just not adequate in many settings. For instance, ski resorts usually take in most of their revenues in 14 weeks. Unable to find an accounting system to meet its needs, Ski, Ltd., developed its own that provides a detailed picture of each resort's revenues and costs on a weekly and sometimes daily basis. Ski's CFO commented:

> We only have 20 weeks or so in which to make our living and finding out in the middle of February how we did in January just wasn't giving us enough time to react. [4]

Each Tuesday, the managers can go over the figures up through the previous Sunday to spot trends and make adjustments. This flexibility has bolstered their financial results.

INFORMATION SYSTEMS

The biggest problem with information systems is that most of the capability is not used strategically. Few organizations have begun to tap the potential of technology to improve effectiveness (as opposed to efficiency). Most management information systems wallow in a tangle of internal information: customer account codes, billing, payroll, and so on, but few use its power to deliver better service to the customer.

One interesting exception is Ski, Ltd., mentioned above. It has enjoyed 30 consecutive years of profit growth and has successfully withstood the demographic, economic, and weather changes in the 1990s while many competitors sank into red ink. It has accomplished this feat through a strategic use of information. Here are a couple of Ski's innovations:

- In a control room, two people monitor sensors and control snow-making equipment, ensuring the most efficient use of the resources. Without this equipment, Ski doubts it would have had a full week of skiing in any of the last four years.
- Its customer information database culls information from a variety of sources so its direct mail campaigns can be tightly focused.
- A computerized staffing system tracks employee movement through the use of bar-coded identification cards and keypads. As skiers flock from one area of the resort to another, managers can glance at the computer to see which employees could be quickly reassigned to handle the overload. As a bonus, this system also tracks hours worked, eliminating payroll paperwork.

In addition to using information technology to manage service delivery, the technology can also be applied vertically toward suppliers and customers. For instance, Kmart supports its highly decentralized merchandising decisions by using a sophisticated information management system and satellite links. Sales data are collected from each store daily and, by 7:30 the following morning, the merchandise buyers can review sales by region and replenish inventories in the distribution centers. Its Partners in Merchandise Flow Program links key suppliers, such as 3M and Johnson Wax, to its inventory databases, letting its suppliers manage their inventories so they don't run out of product.

Kmart also is applying information technology within the stores. It is investigating the use of electronic shelf tags to eliminate the laborious process of retagging merchandise. Its ShopperTrak system, which tracks the number of customers as they enter and move through the store, also is being piloted. This system not only helps Kmart move its sales clerks to where the customers are, it also helps them evaluate the effectiveness of their promotions, because they can see whether shoppers flock to aisles where advertised specials are stocked. These innovations have paid off for Kmart, supporting its five-year effort to transform the organization. The year 1991 was a record-breaking one for Kmart during otherwise tough times, with net earnings rising by 13.6 percent from the previous year and revenue per square foot of floor space rising handsomely. [5]

PLANNING SYSTEMS

Most planning systems are based on a management by objectives (MBO) system of cascading goals. Executives set strategic long-range goals; managers align their shorter-term operational goals to the strategic goals; employees align their individual goals to the operational goals. Who is responsible for thinking in this system? The managers. Who is responsible for doing the work? The employees. This certainly has the earmarks of a traditional paradigm. There are two main problems with traditional planning systems: lack of employee involvement and lack of integration with related systems.

Lack of Involvement

Strategic planning (long-range planning in which strategic choices are made) has long been considered the responsibility of executives, yet few have the same feel for technology, customers, and competitors that frontline employees do. Our world has become so complex that setting strategy is like assembling a puzzle. Relevant information is dispersed among a wide number of individuals. Only when an organization assembles a critical mass of people will it see the whole picture. Breakthroughs and major innovations often require the integration of many smaller advances. In this setting, diversity can be a strategic asset.

Not only do executives frequently lack all the information appropriate to establishing a strategic plan, the exclusivity of the process prevents opportunities to build a common vision that would lead to all employees owning the strategy. Even executives who see the value in involving more people in setting strategy often are stumped by logistics. How can you get involvement within a company of 100 employees, 1,000 employees, or 100,000 employees?

Marvin Weisbord's "future search conference" provides an excellent method for developing commitment to a shared vision between all of an organization's stakeholders. A future search conference typically lasts two days and can accommodate large numbers of people representing all critical stakeholder groups (including employees, customers, suppliers, regulators, and so on). In these conferences, participants forge a shared vision of the future and select initiatives for which they have significant energy to pursue. Mutual understanding and alliances develop from these intense sessions. The outcomes of future search conferences closely align with what one would hope from a collective strategic planning process while permitting wide participation. New rituals, like search conferences, must be designed for organizations to leverage the collective learning of their employees.

Lack of Integration

Planning is a process involving multiple organizational systems: Strategic planning, operational or business planning, budgeting, team goal-setting, measurement, appraisal, and compensation. However, these systems are only loosely coupled and often inconsistent. For instance, the link between the strategic plan and business plans is often only creative writing. The relationship between corporate goals, performance appraisals, and compensation is often even more obscure. Some compensation studies have even found an *inverse* relationship between performance ratings and compensation! Organizations' values scream teamwork while they appraise and reward individuals. Organizations ask for employee input but have no formal way for employees to influence strategy. Organizations tell their employees they are trusted and empowered but do not permit them to sign a purchase order for a box of paper clips. When organizational practices and systems conflict with management's words, employees are quick to notice the discrepancy.

Instead, the information should flow seamlessly across these independent systems and each system should reinforce the same values. The following case study shows how to integrate these systems. [6]

GETTING PEOPLE TO BE PARTNERS IN THE BUSINESS—A CASE STUDY

In the late 1980s, a major automobile manufacturer was implementing self-directed teams within a division. One of the stated values of the organization was teamwork, and corporate communication repeatedly stressed the desire to "invite all employees to be partners in the business." However, when the normal business planning cycle came around, it became apparent that there was no formal way for employees to have input into the business plan. Figure 12–1 shows how their existing performance management system operated.

FIGURE 12–1
Performance System—"As Is"

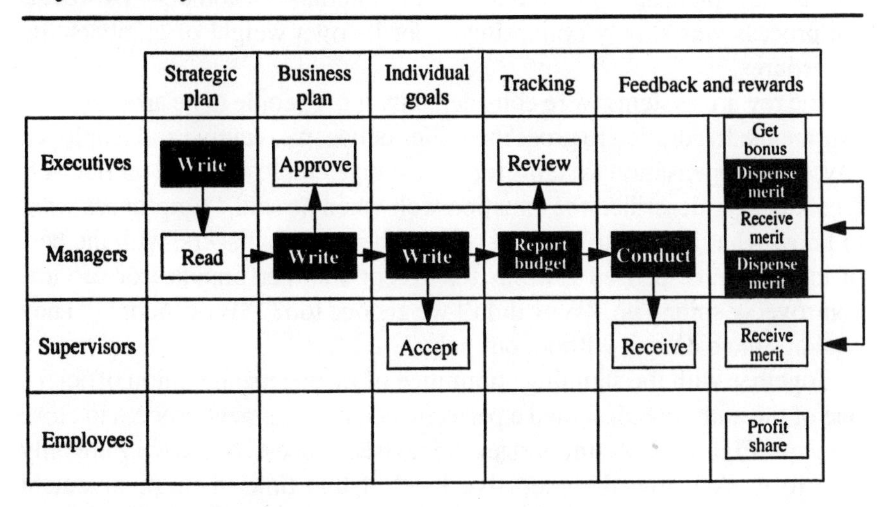

Note: Shaded boxes represent who was primarily responsible for the step.

The strategic plan was communicated by the executive group to division leaders. In a traditional MBO process, managers then wrote business plans and budgets that were tightly linked to the strategic plan. The business plans were written at quite high levels of the organization with no input from those below. In some cases, managers with as many as 20 employees were handed their business plan and budget.

These business plans and budgets were rarely if ever communicated to those below. Employees were simply told what to do. However, they were eligible for profit sharing based on the performance of a system over which they had virtually no control. Compensation experts have estimated that organizations waste 50 percent of their compensation dollars, and this is a perfect example. [7]

There were few meaningful measurements and no reporting requirements below the managerial level. Division leaders watched the budgets and manipulated the business plan goals throughout the year; but this process remained a mystery to most supervisors, union officials, and employees.

The feedback process was just as dismal. Roughly 80 percent of the workforce were UAW members and were not included in any formal feedback process and rarely if ever got informal feedback from co-workers or supervisors. The remaining 20 percent were covered by a progressive process that eliminated ratings, emphasized development, linked loosely to the business plan, and gathered input from internal "customers." However, the process was slowly collapsing under its own weight of cumbersome procedures.

The reward systems were considered by most people to be arbitrary and capricious. In addition to gross inequities between executive and employee pay, the compensation systems for the average employee were ineffective. One manager described the situation well when he said, "Employees want to know what it takes to ring the bell." Whenever managers, with the best of intentions, dispensed rewards for exceptional performance or service, controversy sprang up. "Why didn't we get one too?" "Why them?" "They couldn't have done it without our help."

Together with the standing committee of managers and union officials, one of the authors redesigned a performance management process to close the gaps. Figure 12–2 summarizes the revised process. Since we were only able to work below the executive level at this time, their involvement remained unchanged.

First, the division manager briefed all employees on the strategic plan shortly after he received the information. The management team then identified the broad issues and goals for its division that linked to strategy. Managers then met with each team to explain the strategic issues and broad business goals as well as the budgeting process. Instead of managers writing the business plan, the teams were asked to write their own plans in which they were to establish goals and define any extraordinary budget items. The team business plans and budget items were rolled *up* into departmental and division business plans, which then were subjected to the normal approval process up the chain of command.

Next, teams were provided a team performance log, which helped them track and report on their performance. We established measurements that were directly related to the strategic and business plans on which all teams were to report. Required measures included a customer satisfaction rating, progress toward business plan goals, percentage of self-directed responsibilities the teams had assumed, and so on. Teams that worked on long-term projects were required to report on performance to budget and schedule. Many teams came up with other measurements they wanted to track, such as equipment downtime (to help justify new tools and equipment). Teams received assistance on measuring, tracking, and graphing these indicators and were expected to report quarterly on their results.

FIGURE 12–2
Performance System—Revised

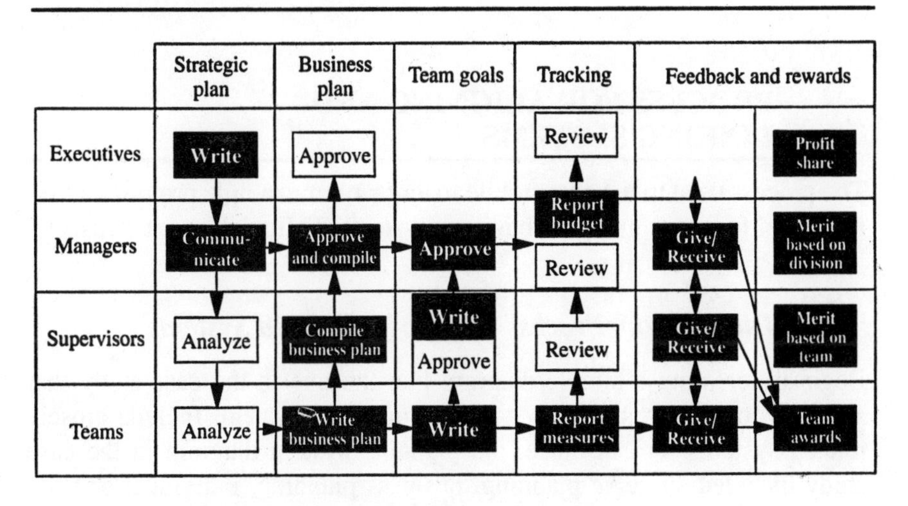

Note: Shaded boxes represent who was primarily responsible for the step.

Teams conducted their own performance reviews in conjunction with the planning process, using our open appraisal format (see Chapter Seven). As a group, they assessed individual and team performance and set new team and productivity related goals quarterly. Supervisors participated in the feedback process.

The reward systems were still in flux. Our recommendation was that they adapt compensation systems already in place so teams could receive both monetary rewards for bottom-line improvements and nonmonetary rewards for extraordinary performance through nomination by other teams, customers, or managers.

Was the time spent involving front-line employees worth the effort? Even managers who had felt the employees were not ready for this involvement were impressed. Although only a one-year plan had been requested, the powerhouse team built a five-year business plan to decommission its aging power plant and replace the power with stand-alone units, redeploying the team members into more meaningful roles. (Had management suggested this, the union undoubtedly would have filed numerous grievances.) Another department decided to become a profit center, turning an expensive hazardous waste stream into a recycled, salable product. Other teams asked for small items to improve their productivity, which managers quickly approved under the current year's budget.

PROCESS FOR ALIGNING AND LINKING SYSTEMS

The process used in the automobile manufacturer example provides a useful model for redesigning all business systems. Follow the next steps.

Diagram the Process along with Related Systems

Begin by creating a simple diagram like the one in the case study that shows who performs the key tasks within each system. Include closely related systems. For instance, the planning system diagram in the case study included strategic planning, business planning, individual or team goal setting, and performance appraisal and compensation, because these systems are directly involved in the process of planning or should be intimately linked to the plans.

In our experience, this step often reveals significant ignorance about how the systems interrelate, so creating this diagram may represent an important step in learning. Sometimes it is helpful to create two diagrams: how the procedures say the systems work and how they actually are carried out. The disparity may point to important issues that need to be addressed.

Analyze the Information

Next, analyze the diagram, looking for errors in involvement. This usually is manifested as people left out or as the wrong people in control. Frequently, we find that those who must carry out the decisions or who are most closely linked to critical knowledge (e.g., those who work directly with customers) are not involved. Make sure that the key knowledgeholders and stakeholders are represented and involved.

Then, look for information and feedback loops that are absent or going the wrong way. For instance, managers routinely appraise employees; but the reverse is rarely practiced, while both manager and employee alike need feedback.

Watch for inconsistencies in system values. Examine each system to see if it supports the values of quality and empowerment. As Michael LeBoeuf, author and professor at the University of New Orleans, states, the "greatest single obstacle to the success of today's organizations is the giant mismatch between the behavior we need and the behavior we reward." [8] In this context, don't confuse rewards with compensation. Rewards also include public recognition, involvement, an opportunity to reduce the hassle necessary to do work, and so on.

Last, but certainly not least, look for alignment to customer needs. Does the process solicit and act on emerging customer needs? Do the mission and values focus all levels on meeting and exceeding customer expectations?

Redesign the Systems

Sometimes, the fixes are easy. Perhaps changing the direction of information or adding a source of feedback is all that is necessary. Occasionally, new methods, such as future search conferences, exist that provide useful models. However, most often, a new system must be invented. Begin by defining the outcomes for the system and the values it should support. Then be willing to relinquish old artifacts in favor of more appropriate methods. No one knows what the organizational systems of the future will look like; we just know that the organizational systems of the past are obsolete. Innovation and vision must guide your efforts.

CONCLUSION

Organizational systems have not kept pace with other changes within the workplace. By reinforcing the old competitive standards, accounting systems often yield poor management decisions, information systems wallow in bureaucratic data, and planning systems preclude participation from all levels. For teamwork and quality to become a reality, these systems must be reinvented. No one can provide a formula for success; but, in this chapter, we have provided numerous examples of how organizations have changed these systems to be more customer-focused. These examples provide a glimpse of the extensive transformation required to achieve total quality.

Tips, Tools, and Techniques
A SYSTEMS "HIT LIST"

Here is a partial list of systems that should be reviewed for alignment to total quality principles.

Financial systems

- Accounting and finance.
- Budgeting.
- Purchasing.
- Stock issue and ownership.

Human resource systems

- Benefits.
- Compensation and rewards.
- Hiring and selection.
- Performance appraisal.
- Position descriptions and job evaluation.
- Training.

Information systems

- Benchmarking.
- Customer and supplier feedback.
- Measurements.
- Reporting.

Planning systems

- Strategic planning.
- Business or operational planning.
- Production scheduling/work planning.
- Inventory management.

SUGGESTED READING

Johnson, H. Thomas. *Relevance Lost: Rise and Fall of Management Accounting.* Cambridge, MA: Harvard Business, 1991.

———. *Relevance Regained: From Top-Down Control to Bottom-Up Empowerment.* New York: Free Press, 1992.

Lawler, Edward III. *Strategic Pay: Aligning Organizational Strategies and Pay Systems.* San Francisco: Jossey-Bass, 1990.

Osborne, David, and Ted Gaebler. *Reinventing Government: How the Entrepreneurial Spirit Is Transforming the Public Sector.* New York: Addison-Wesley Publishing, 1992.

Weisbord, Marvin. *Productive Workplaces: Organizing and Managing for Dignity, Meaning and Community.* San Francisco: Jossey-Bass, 1987.

———.*Discovering Common Ground: How Future Search Conferences Bring People Together to Achieve Breakthrough Innovation, Empowerment, Shared Vision, and Collaborative Action.* San Francisco: Berrett-Koehler Publishers, 1993.

Chapter Thirteen

Organizational Learning

How to Leverage Knowledge and Experience

Often TQM is mistaken for a goal, an end state, when in fact it is a process for *continuous* improvement and innovation. To be able to improve and innovate continuously requires the ability to capitalize on experience and to leverage what has been learned. Few organizations seem to have made the connection between learning and total quality. This chapter examines the obstacles that inhibit learning in organizations and suggests strategies for creating a continuous learning environment.

The headlines of the last decade have heralded this as the Information Age. Information is lauded as the new currency, the raw material that feeds every enterprise. While information is vitally important to the success of any organization, it is, in itself, insufficient. Consider it from a historical perspective.

In the early 50s, the competitive advantage for most businesses was based on the ability to out "do" the competition. The most successful companies were able to produce more goods and services faster and cheaper than anyone else. American industry was, and still is, the champion in this sport. During the 70s and 80s, however, the edge began to go to those organizations that could leverage and operationalize information. Knowing more than the competition gave companies the edge. Today, because information is so accessible, plentiful and ever-changing, it is impossible to maintain the exclusivity of knowledge. Thus, today's advantage comes from being able to out "learn" the competition: to improve and learn continuously. This age, then, might more accurately be called the age of the "Learning Organization."

What is a learning organization and how does continuous learning differ from knowing or being able to leverage knowledge? Best-selling author Peter Senge explains it as the difference between adaptive learning and generative learning. [1] Adaptive learning is what enables organizations to accommodate the accelerated rate of change associated with our times. It is a minimum requirement for staying in business in the 90s. An adaptive organization is flexible and responsive to the ever-changing needs of its market. To be truly competitive, however, organizations also must practice generative learning. Generative learning refers to the ability to expand capabilities and create something *new* out of given information or knowledge.

While there is no simple recipe or formula for becoming a learning organization, there is a common set of characteristics that all learning organizations share:

- All organizational members are aligned to a shared vision. This is the first step toward continuous learning. It is important because it ensures that all knowledge creation and collection is done with the collective purpose of the organization in mind.
- Learning occurs spontaneously, often outside the bounds of traditional instructional settings.
- The organization recognizes that experimentation and risk taking are integral to learning.
- Management not only demonstrates support for learning but models learning as well.
- Group learning is emphasized over individual achievement. Random, uncoordinated learning by individuals has no significant impact on the effectiveness of the organization.

OBSTACLES TO LEARNING

Psychologists tell us that the drive to learn is one of the strongest innate human urges—stronger, even, than our sexual urges. Yet we see little learning behavior occurring in our organizations. Ironically, organizations traditionally have created environments that not only stifle the natural urge to learn but actually work energetically to counteract it.

Knowledge as Status

Some of the most time-honored cultural traditions are the biggest culprits in discouraging organizational learning. Our working environments perpetuate those values learned in our school system. They are the attitudes that value being "right" above all else; that equate admitting we don't know with incompetence; that link knowledge with power, status, and reward.

Mistakes are to be avoided at all costs because they are entered permanently on your scorecard and have the cumulative effect of "bringing down your overall average." We learn to suppress our own doubts and hide our ignorance lest we jeopardize our position in the world. When practiced in our organizations these values effectively discourage learning.

The learning organization, on the other hand, encourages self-doubt and fosters risk taking. Mistakes are viewed as valuable learning opportunities. Uncertainty and instability are seen as useful because they tend to yield innovations and creative ideas. Competitive companies recognize that a sure thing holds little promise as an advantage, and they willingly suffer the confusion and uncertainty that goes with being on the cutting edge.

Myths about Learning

We also are hampered in our efforts to foster continuous learning by our misunderstanding of how learning occurs. For most people the term *learning* conjures up passive, episodic events in which someone else—an expert—pours knowledge and information into our brains. This approach does not leverage our innate desire to learn. Rather, it reinforces the notion that we are not capable of learning by ourselves. This approach turns out dullards who have had the curiosity and the ability to think for themselves beaten out of them. In the end we stop trying to aspire to levels beyond those expected by our teachers.

Our notion that learning is limited to intellectual understanding is also a handicap to fostering continuous learning. This narrowly focused definition misses the most vital component of the learning process: the internalization of our understanding and the new ideas and action that result from it. How many times have we seen managers nod enthusiastically in agreement with the principles of empowerment and then turn around and execute their duties like benevolent dictators? They have all had the training, they can recite the concepts perfectly—but they have failed to translate the knowledge into action. Learning is incomplete until new understandings have been integrated and become habits of behavior.

FIRST STEPS TO BECOMING A LEARNING ORGANIZATION

The path to creating the learning organization frequently starts with the most obvious steps. Organizations in this country are slowly waking to the value and potential high return of developing their human resources through training. Research conducted by the American Society for Training and Development in conjunction with the Department of Labor revealed that formal schooling and job-related training consistently have proven to have greater impact on improving productivity than have investments in machine capital. Training and education contributed about 81 percent to all improvements in the nation's productive capacity, while machine capital contributed only about 20 percent. [2]

In spite of this data, the same study indicates that organizations in the United States currently spend only 1.4 percent of payroll on training. Their foreign counterparts operating in the United States, however, average 5 to 7 percent of payroll on training, as do some of America's most progressive and competitive companies. Several Malcolm Baldrige Award winners, Motorola and Xerox among them, each spend millions per year on training and have built company universities dedicated to the continuing education of their employees. Private industry is not leading this trend alone. Public institutions are making similar commitments. The state of Oregon, for example, has established learning benchmarks for all state agencies and departments that set a goal for 20 or more hours of training for each employee every year.

BEYOND TRAINING

As with TQM, however, true organizational learning requires more than implementing a few isolated programs. Learning needs to become integrated into the fabric of the organization, internalized into its way of life. John Coné, director of Sequent University for Sequent Computer Systems in Beaverton, Oregon, and previously with Motorola, has a clear vision of an integrated, continuous learning environment. As he describes it, learning events in a learning organization "are less than one minute in length and take place within 20 minutes of the learning need having been identified." [3] As simple as it sounds, achieving this kind of integrated, continuous learning requires addressing the issue on three different fronts: at the organizational level, the leadership level, and the team level (see Figure 13–1).

ORGANIZATIONAL STRATEGIES

In earlier chapters we have made a case for reexamining the organization structures and systems to assure that they are in alignment with quality efforts. Appropriately aligned systems and structures are just as critical to

FIGURE 13–1
Creating the Learning Organization on Three Fronts

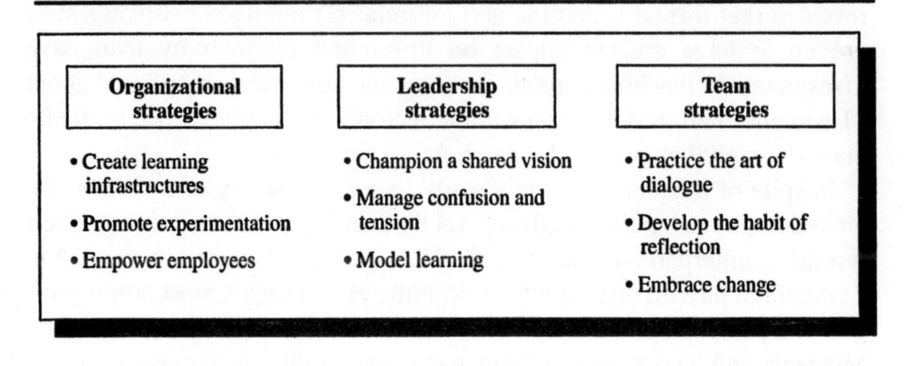

creating a learning atmosphere. Here we describe three strategies for creating an organizational culture that supports learning. The first strategy suggests establishing a supporting infrastructure much in the same way that organizations do for other practices like training. The second addresses methods for promoting experimentation and exploration, and the third deals with empowering employees.

Organizational strategies
• Create learning infrastructures
• Promote experimentation
• Empower employees
Leadership strategies
• Champion a shared vision
• Manage confusion and tension
• Model learning
Team strategies
• Practice the art of dialogue
• Develop the habit of reflection
• Embrace change

Create Learning Infrastructures

For most organizations, expanding training's role is a logical place to begin building a learning organization, principally because there is a workable infrastructure already in place. To reach out from the training room to the work setting where the best learning takes place, however, will require establishing equally effective supporting "learning" infrastructures.

Any learning infrastructure should, at heart, maximize knowledge sharing. Suggestion boxes, for example, are a time-honored system for soliciting ideas from employees. They have the right intention; but, because they tend to hide information from public view, they actually work counter to the learning process. Learning only occurs when ideas are shared, not hoarded. Chaparral Steel, a company that has successfully survived the near death of an industry, is proud that it can seldom identify any single person as the source of a good idea because its procedures for managing information and innovation are so open and collaborative. [4]

Sometimes good learning infrastructures are created in the simplest ways, such as arranging work spaces to allow for accidental meetings among employees, or directing traffic flow to encourage frequent contact. The informal chance meetings that these structures create frequently provide the most productive exchanges of information and feedback. Another example of open knowledge-sharing structures are fishbowl-type problem-solving sessions. These are open meetings where other organizational members are invited to actually watch teams at work. At G.E. these are called "work-out sessions" and are an important part of the learning infrastructure.

Another useful strategy involves extending the systems or practices that already are used in training to reach back to the job level and encourage application and experimentation with new skills. These strategies may include training techniques as simple as required outside assignments or follow up with participants after training encounters. Motorola successfully expanded its training infrastructure and encouraged on-the-job learning by creating what it calls "applications consultants." These employees receive special training on a topic or skill and then act as local technical consultants to co-workers. This strategy facilitates just-in-time learning and provides immediate solutions for real problems.

Promote Experimentation

Human curiosity compels us to explore and learn. Left to our own devices, we naturally seek out information in response to our own needs and questions. Organizations that value exploration and encourage risk taking send a strong message to employees that new ideas are valuable, even if they don't directly generate a profitable product or process.

Chaparral Steel, which embodies this attitude, has a companywide motto that states simply, "If you have an idea, try it." [5] Honda sponsors an annual "idea contest," for which the company funds experimental projects generated by employees. [6] Minnesota has instituted a program called STEP (Strive Toward Excellence in Performance), in which a board appointed by the governor reviews and selects for implementation innovative proposals submitted by teams of state employees. [7]

How does an attitude like this become embedded in the culture of an organization? John Coné tells a story of a policy instituted during his tenure at Motorola that unexpectedly achieved this kind of cultural shift. Early in its quest for continuous learning, Motorola initiated a policy of mandating that all department managers dedicate a certain portion of their budget to training. In the beginning, few of the managers took advantage of the potential of this policy. Some used the money rather thoughtlessly, believing it was better to spend it on anything than to lose it. Over time they began to realize the potential value of using training to fulfill business needs and became more vocal and proactive about their needs as well as more involved in the creation and delivery of the service.

This was a valuable outcome in itself, but the evolution didn't stop there. The managers began to get more creative. They wanted to know if they could get credit for nontraditional learning events that they created themselves? What if one department employee trained a co-worker on the job? What about a team of workers studying a new problem? What about a new procedure for sharing and transmitting information among employees? Motorola's response was an enthusiastic "of course!" These are exactly the kind of activities that, when leveraged and shared, begin to build virtual learning environments.

Empower Employees

Perhaps the biggest organizational obstacles to learning are the power structures and hierarchies that assign varying degrees of authority to individuals in the organization. In traditional work environments, knowledge is equated with power, and sharing knowledge diminishes power. Undoing these strong cultural notions requires a fundamental shift in our power paradigm. As Marvin Weisbord said, "It takes wisdom to see that no one person knows enough to do it all, and courage to involve those who have the information and control the change." [8] This requires going beyond creating teams and giving them responsibility for new tasks. It requires freely sharing information with teams, empowering them to use it, and crediting them with the ability to benefit the organization. We discussed this issue in more detail in Chapter Nine.

Without empowerment we simply reinforce the old learning paradigm that we need to be taught by some credible "expert." In reality, people learn more effectively when they have a personal commitment to learning, and when they think that their learning will have consequence. According to Peter Senge:

> People learn most rapidly when they have a genuine sense of responsibility for their actions. Helplessness, the belief that we cannot influence the circumstances under which we live, undermines the incentive to learn, as does the

> belief that someone somewhere else dictates our actions. Conversely, if we know our fate is in our own hands, our learning matters. [9]

Southwest Washington Medical Center is seeing the benefits of empowering employees to learn and make critical business decisions. In an environment with a strong historical link between knowledge and status, the hospital has successfully created cross-functional patient care teams. The program, called PartNURShip™ combines nurses, technical specialists, therapists, and caregivers on one team that focuses on a specified caseload of patients. [10] Before the teams were formed, these roles were separated by strict functional barriers. Now, not only does the patient receive more personalized and higher quality care from a consistent team of professionals, but the team members learn and act with a more holistic approach and understand better each other's roles and the interdependencies among them. This kind of cross-functional learning could never have happened in the old environment.

LEADERSHIP STRATEGIES

As in most organizational efforts, the bulk of responsibility for fostering a learning organization falls on the shoulders of its managers and leaders. It is the managers who not only create and manage the organizational systems that facilitate learning but also manage the interpersonal relationships that empower and encourage learning in employees. To be most effective in this role, leaders need to fully adopt the management style advocated in Chapter Ten. In addition, they need to perform three critical tasks: champion a shared vision, manage confusion and the tension it creates, and model learning for all employees.

Champion a Shared Vision

A clearly articulated and shared vision is critical to the creation of a learning organization. It must be clear so that all efforts are aligned, and valuable contributions can be distinguished from fruitless ones. It must be shared so that it inspires the commitment and energy of all employees. As Peter Senge explains:

> Shared vision is vital for the learning organization because it provides the focus and energy for learning. While adaptive learning is possible without vision, generative learning occurs only when people are striving to accomplish something

> that matters deeply to them. In fact, the whole idea of generative learning—expanding your ability to create—will seem abstract and meaningless until people become excited about some vision they truly want to accomplish. [11]

The leaders of Southwest Washington Medical Center realized the importance of a shared inspiring vision. Instead of drafting its vision in the seclusion of the boardroom and then trying to sell this vision to employees, it took a much more participative approach. All employees were asked for their answers to two questions, Where are we going? and How will we get there? The responses were used to create a four-point organizational vision:

- Foster a learning organization.
- Develop and integrate health systems.
- Become community focused.
- Become patient centered.

The hospital published the four points on posters, which also included almost 400 of the actual employee responses, and distributed them throughout the hospital.

Manage Confusion

Confusion is an uncomfortable state for most of us, yet it has a good deal of value to learning. Confusion usually occurs when an individual or group is faced with a situation that requires more than their current experience or base of knowledge can handle. It's like having a map that does not match the territory. The discomfort caused by confusion triggers an immediate search for meaning to reduce anxiety. This initiates the learning process and presents opportunities for creating something fresh and innovative.

Good leaders know how to manage and leverage confusion effectively. If they allow confusion or creative tension to become too great, they will frustrate or burn out workers. If tension or confusion is too low, then there is no perceived need to learn or change. Comfort and stability tend to create a false sense of security, blinding us to the need for continuous improvement. The optimum discomfort level, on the other hand, promotes continuous learning and innovation. Keeping workers slightly out of their comfort zone, but still motivated to improve, requires celebrating past and current successes continuously while constantly championing the vision and turning successes into new challenges.

In a *Wall Street Journal* article Peter Drucker described a Japanese strategy of systematic abandonment of products. [12] While still celebrating the successful launch of a new product, many Japanese firms set new deadlines for replacing that product with another one. The new product development effort pursues three approaches simultaneously: incrementally improving the existing product, developing a new product from the existing one (something they call "leaping"), and pursuing an entirely new innovation. This practice discourages complacency and ensures that groups continue to learn and innovate.

Model Learning

Peter Drucker once said, "A superior who works on his own development sets an almost irresistible example." [13] Actions speak volumes more than words. Leaders who pursue their own learning send a much more powerful message than when they teach. Some organizations take this very much to heart. Ed Simon, president and COO of Herman Miller, requests that each of his senior managers devote as much as 25 percent of work hours learning to become what he calls "organizational architects." [14]

Modeling learning behaviors, however, can mean adopting some habits that may seem risky. Good learners, for example, freely admit what they do not know. They acknowledge their mistakes and publicly talk about what they have learned from them. They acknowledge that someone of lesser rank may have something to teach. They approach every situation as a learning situation and every person as someone from whom they can learn.

In his book *The Learning Edge,* Calhoun Wick relates a story about a CEO and one of his managers who were touring a plant to inspect some equipment they were considering purchasing. [15] The manager summed up the condition of the equipment very quickly and determined that it was not a good buy. The CEO, on the other hand, climbed all over the machines, took copious notes, asked questions of operators, and checked sizes of motors. Upon comparing their experiences, Wick found that the manager had only a superficial knowledge of the equipment. The CEO, however, had through his investigation learned a great deal about what it took to run the operation as well as some of the factors that contributed to its success.

How differently the two approached the experience and how effectively the CEO modeled good learning behavior!

TEAM STRATEGIES

When we talk about learning organizations, we really mean the collective learning of the individuals within the organization. The American Society for Training and Development's seminal study on fundamental workplace skills, *Workplace Basics: The Skills Employers Want,* places at the top of its list the ability to "know how to learn." [16] Without this ability, other meaningful development is rarely possible. But true *organizational* learning only occurs when the learning is collective, when workers communally learn and act on their shared learning. The skills for achieving this kind of learning are not taught in school. While many traditional learning and study techniques may be effective for individual learning, they do not help us with the collaborative learning strategies that lead to new understandings and actions. Collaborative learning calls for skills that are foreign to our culture. This section focuses on three of the skills that, when used together, promote collective learning and collaborative action: the art of dialogue, the habit of reflection, and the ability to embrace change.

Practice the Art of Dialogue

In his book *The Fifth Discipline,* Senge makes a case for the importance of dialogue and discussion to team learning. Dialogue is the "free and creative exploration of complex subtle issues, a deep 'listening' to one another and suspending of one's own views." Discussion, on the other hand, is when "different views are presented and defended and there is a search for the best view to support decisions that must be made." [17] One is necessary for learning and innovating; the second is critical to decision making and action. It is not a surprise that in this action-oriented and results-oriented culture we excel at discussion but are sadly inept at dialogue.

Dialogue as a learning strategy allows people to gain new insights that they would not have reached working alone. Dialogues are divergent conversations aimed at reaching a new and richer understanding of an issue or problem. It is only possible in groups where trust is high and authority and hierarchy are absent. It requires recognizing our assumptions, distinguishing them from those thoughts or beliefs actually based on facts, and then holding them up for critical public examination.

Many organizations make dialoguing a regular habit. Hanover Insurance holds biannual retreats for its management staff. The agendas for these retreats are purposely kept loose. There is no attempt at decision making. The time is simply dedicated to coming to those richer understandings that help make for better decisions. Mike Warn, of Warn Industries, who has a long-standing reputation in the Northwest for participative management practices, years ago instituted monthly communication meetings with employees. These sessions are open to 11 employees at a time. They are scheduled by the employees' birth month to ensure that participants represent a cross-section of the organization. During these meetings employees have the opportunity to dialogue with the president and with each other on any issue, question, problem, or rumor that may be on their minds. Not only has it proven a valuable means for the president to stay in touch with employee concerns, it has also been a useful vehicle for sharing information, modeling learning, and reaching a deeper understanding among employees.

Develop the Habit of Reflection

It seems ridiculously self-evident to say that thinking has value as a learning strategy. Yet we don't seem to have the skills for reflecting on our own processes and experiences. For most of us, reflection connotes something mystical. It brings to mind the image of a meditating monk cloistered in silence. Taking time to think, and to reflect on our own thinking, is generally not valued in our action-oriented world. In our culture we call it "daydreaming" and consider it an idle occupation. Consider the observation of one manager who had worked both in Japan and the United States. In the United States, she noted, if we encounter someone sitting in his office staring out a window, we assume he isn't doing anything important and can be interrupted. If we encounter him up and moving around, performing some tangible task, we infer that he is busy and should not be disturbed. In Japan just the opposite is true. No one would presume to interrupt someone who was thinking.

Americans are famous for our "ready, fire, aim" approach to situations. Peter Senge describes a series of experiments he conducted with managers who were working with a computer simulation. [18] Invariably, they would choose a course of action, abandon it when it showed signs of failing, and then immediately adopt another strategy. Even when time was plentiful, they consistently failed to do any analysis of why their strategies failed or what they might have done differently to make the strategies work. Reflection is the antidote to this haphazard approach to problem solving. Taking time to think, to monitor our own cognitive processes, and to leverage those strategies that work makes us conscious and efficient learners. As Chris Argyris argues, until people learn to reflect critically on their own behaviors and understand their own cognitive rules or reasoning (what he calls "double loop learning"), we are not truly learning. [19]

The habit of reflection also can make us aware of the assumptions on which our behavior is based. These assumptions are our untested generalizations about the world that frequently stand between us and our ability to dialogue and learn. Until we are able to examine and confront our own beliefs, we may be blocked from hearing and integrating conflicting views.

Several strategies are useful for developing the habit of reflection. Journals, for example, are effective tools for developing reflective thinking. Individuals who document their activities and thoughts frequently are surprised by what they learn about themselves. Journals also help them remember those actions that were effective and those that were not and help prevent them from repeating failures. Since it is collective learning that nets the highest return for organizations, teams should be encouraged to keep journals as well. Team learning is more than the sum of the learning of its individual members, so a journal or record can document for all the new understandings and ideas that the group has generated together.

Another strategy for teaching reflection was developed by Chris Argyris of MIT. The activity, which he calls the "left hand column," asks participants to recall some recent, significant interaction. [20] On the right hand side of their page they record what was actually said. On the left hand side they record what they were thinking, but didn't say. The exercise forces participants to examine their own processes and bring to light those assumptions that drive their behavior and impact their ability to hear and learn.

Embrace Change

Employees in the learning organization must understand that they have a fundamentally different kind of employment contract with employers. People are no longer hired so much for what they know but for what they can learn. The skills and knowledge new employees bring to the job no longer assure them of a lengthy career and even may be irrelevant before their probationary period ends. Employees are not only expected to learn and adapt to new roles but to embrace the challenge of change and be proactive and self-directed about their own learning. At Sequent Computer Systems, for example, employees are not expected to wait for the organization to identify changes they need to make—they are expected to develop their own interests that relate to the business and propose new alternatives.

Tom Peters makes a case for a whole new set of selection criteria that fly in the face of our traditional notions. He recommends weeding out the dullards and focusing on "curious people," making sure to include a few truly "off the wall" people when hiring or promoting. [21] Bright Wood Corporation, a wood products manufacturing company in Oregon, puts a great deal of energy into its selection process. Recognizing the volatility of its industry, its selection criteria put adaptability before experience. The company, which currently employs about 600 people, processes 2,000 to 3,000 employment applications each year. Each applicant undergoes an average of five interviews and screening activities before being hired. The activities include role playing and group problem solving to identify the applicant's facility and comfort with adapting and innovating. As a result of this careful screening, Bright Wood has one of the lowest turnover rates in the business and is thriving in an industry fraught with competitive and environmental pressures.

CONCLUSION

Organizations that continuously promote and leverage learning are in the best position to truly integrate and live total quality management. That human beings are naturally inclined to learn and explore is a resource that can work tremendously to the advantage of any organization. However, undoing the damage we have done by thwarting this natural inclination will require some fundamental restructuring of the way we do business. Those who are willing to expend the effort and time, to examine and realign their practices at the organizational, leadership, and team level, can realize a potentially enormous payoff. As Citibank CEO Walter Wriston predicts, "The person who figures out how to harness the collective genius of the people in his or her organization is going to blow the competition out of the water." [22]

Tips, Tools, and Techniques
TEAM LEARNING STRATEGIES

Conducting Dialogues

A true dialogue allows team members to explore options and ideas that go beyond what they could invent individually. To be effective, however, a dialogue demands that each participant agree to follow these rules:

- Allow each other equal opportunity to speak.
- Interrupt no one at any time.
- Listen attentively.
- Accept everything that is said as true and possible.
- Focus on expressing beliefs, needs, and hopes, rather than justifying a position. [23]

After everyone has had the opportunity to speak, the team should take some time to invent solutions or options that include as many as possible of the hopes and beliefs expressed. The point here is not to find the right or even the best idea so much as to gain new insights and innovations.

Team Journals

Team journals are records of the collective learning of a group. When regularly kept, they help to establish a habit of attending to learning and of valuing it as an achievement in and of itself. Journals may take any form but should at the very minimum be accessible, storable, and easy to maintain. If the effort of keeping a journal becomes too great, it will inhibit its use. This implies leveraging the existing record-keeping systems. Slight modifications to meeting minutes, for example, would make it easy to include key learnings, significant milestones, action results, or dialogue discoveries.

A team journal is a waste of effort, however, if it is not used. Teams should establish a ritual of periodically reviewing them. These reviews should accomplish these functions:

- Assure that the information being logged is useful and relevant.
- Highlight trends, discoveries, or insights.
- Identify lessons learned.
- Establish a means for sharing key learnings with others.

SUGGESTED READING

Argyris, Chris. "Teaching Smart People How to Learn." *Harvard Business Review,* May/June 1991, pp. 99–109.

Bohlin, Ron. "Organizational Learning in Practice: Interview with Ray Stata CEO of Analog Devices, Inc." *The McKinsey Quarterly,* Winter 1992, pp. 70–83.

Leonard-Barton, Dorothy. "The Factory as a Learning Laboratory." *Sloan Management Review,* Fall 1992, pp. 23–38.

Macher, Ken. "Organizations That Learn." *Journal for Quality and Participation,* December 1992, pp. 8–11.

Nevens, Michael. "Organizational Learning in Practice: Interview with Craig Barrett, Executive VP of Intel Corp." *The McKinsey Quarterly,* Winter 1992, pp. 83–87.

Pasmore, William, and Mary Fagans. "Participation, Individual Development, and Organizational Change: A Review and Synthesis." *Journal of Management,* June 1992, pp. 375–99.

Senge, Peter. *The Fifth Discipline.* New York: Doubleday, 1990.

———. "The Learning Organization Made Plain." *Training & Development Journal,* October 1991, pp. 37–44.

Wick, Calhoun, and Lu Stanton Leon. *The Learning Edge.* New York: McGraw-Hill, 1993.

Epilogue

The Future
The Total Quality Community

For a traditional organization, making it to Phase 3, Integration, may seem only a pipe dream. The massive changes to organizational culture and practices can be overwhelming. However, achieving integration is only the entrance requirement to becoming a viable organization in the next century. The rate of change will only increase. The high-speed bullet train is leaving 21st Century Station, and being a total quality organization is the price of a ticket. Product life cycles will run on short tracks, technological advances will arrive on magnetic rails, and new competitors will crowd on board every day. To meet changing customer needs and expectations, the train must travel faster each trip. The once comforting and predictable thwapity-thwap has been replaced by a carnival ride. Only the most adaptable customer-focused organizations will keep from being thrown off around the tight corners ahead.

Where is this train taking us and what will make us successful in this new world? No one can know for sure, but we invite you to ride with us as we explore one possible future: the total quality community. In this chapter, we will explore why we believe a total quality community will be necessary for our nation's competitiveness, what one might look like, and how to begin creating a total quality community.

WHY CARE ABOUT THE COMMUNITY?

In our mobile society, *community* has come to refer most frequently to a geographic boundary. However, in the past, community connoted much more: intimate interdependencies, shared values, close relationships, linked potentiality. We believe that we must return to such communities to maintain competitiveness in the next century.

Each organization exists within a much larger system. Attending to quality solely within a business may not be sufficient if government, education, and other interdependent segments of society are not aligned. For instance, as a nation, we recently have come to understand the impact of health care costs on our competitiveness.

Organizations that are succeeding with total quality have an obligation to carry their learning to other segments of society. Many organizations already are making demands on their suppliers and distributors. This trend will by necessity extend to other segments of our society. In fact, this already has begun in some places around our country. Pensacola, Florida, has brought together people from many sectors to create a "Quality Community." The state of Oregon, through its Benchmarks and Human Investment Strategy, is striving to create a high-performance society. A scattering of communities like Visalia (California), Madison (Wisconsin), and Phoenix (Arizona) are "reinventing government" and establishing mutually profitable partnerships among schools, city governments, and private businesses. [1]

This involvement in external organizations must go beyond the current opportunism. Many organizations now encourage volunteerism and community involvement through the lens of public relations or charity, where the activity is justified based on "looking good" or on moral righteousness. However, involvement of this kind in the larger community should be viewed as strategically important.

Since our organizations' railroad cars are all hooked together and thus affect each other and are affected by each other, this outreach is not charity; it is an investment, necessary for survival. Taking a long-term view, business has as much stake in prenatal care and teen pregnancy rates as social services has in the health of business to provide high-wage jobs. Social and economic policy are two sides of the same coin.

WHAT WOULD A TOTAL QUALITY COMMUNITY LOOK LIKE?

For many decades, we viewed our communities as machines: independent pieces, each with a discrete function. Using reductionist logic, we broke tasks into increasingly smaller pieces, as if dissecting the organism of society would help it live better. We separated learning from work, school from business. Over time, we relinquished our individual responsibility to society. We delegated our will to government in hopes that we would not have to participate in the messy business of self-governance; we delegated our acts of compassion to charitable organizations; and we delegated our children's development to the schools.

Having effectively separated all these sectors of society, we now are distrustful of any that move to influence another. Schools bristle at business influencing their curricula. Private industry carefully avoids entangling itself in social policy for it wants no part in solving social ills. Government views turning a profit as distasteful. But like gears that have been pulled apart, the sectors of our society are disengaged. Stuck in neutral, we waste energy and resources.

As an example, Meadowood is a small business in Oregon that manufactures pressboard from waste grass straw. Its product not only reduces the need to harvest valuable timber, it also reduces the polluting practices of field burning. [2] It can't keep up with demand for its product, however, because of lack of capital. The timber industry, on the other hand, has millions of dollars of idle capital equipment for lack of trees to cut. Meadowood and the timber companies seem like two problems that should meet, yet even within one sector of society, no means other than serendipity currently exists to bring them together.

Instead of thinking of our community sectors as mechanical gears, each with its own separate function, we would do better to use an ecological metaphor. Machines imply simple cause-effect relationships and reductionist thinking. Ecology, on the other hand, implies a systems view where complex interrelationships are present. For example, some community members argue that if they had more and better businesses, their schools would be better; while others argue that if they had better schools, they would have better businesses. In reality, these are not chicken-and-egg arguments. Both positions are correct. They are in a mutually reinforcing cycle—for better or for worse. The trick, then, is to get the cycle moving in the right direction. This will necessitate bringing our whole society into alignment.

One Model—Mondragon Cooperatives

Perhaps the Mondragon Cooperatives in Spain can provide a model for integrating these sectors into an ecologically sound system. [3] These Basque cooperatives combine socialist principles with capitalist structures to manage their community as an integrated system. For instance, their banks become, in effect, partners with their borrowers, charging lower rates for socially useful endeavors and providing additional resources and lower interest rates for struggling ventures. Most of the depositors are Basques, who expect the money to be used to provide "newstreams" and employment in their communities. Employees are also owners of their employing businesses, because roughly one-third of their salary is taken as equity. This unusual merging of social and economic policy has yielded the highest productivity per worker in Spain.

Application of TQM Principles

Alignment of community sectors is important but not sufficient for a total quality community. Each sector also must eliminate waste.

Imagine a society where all sectors applied TQM principles. What would our society be like if government focused on reducing cycle time, if education reduced batch sizes of students so they could progress at their own rate, and if business aligned all its systems to support quality? What if all nonprofit organizations benchmarked their performance against best practices around the world, and colleges turned out graduates in a just-in-time fashion? What if hospitals viewed the patient (not the health care provider) as customer? Envision a six-sigma judicial system. On a more personal scale, imagine filing an annual tax return after the IRS removed all nonvalue-added activities.

Experience has shown that cycle time—the time it takes to manufacture a product or to provide a service—generally can be reduced by 70 percent. Reducing cycle time frees enormous blocks of time and improves flexibility. If our schools could teach the equivalent of a high school education in one-third the time, we would have time to expand student learning. Similar gains could be made in every sector of society.

Empowerment

What could we do with all the saved time and resources? An answer: We could expand empowerment throughout organizations and into the community.

Imagine a society where all citizens are actively involved in framing the direction of their communities, where everyone feels they can and should contribute to society. How much more effective would our schools be if teachers were empowered to act on the changing needs of their students? Imagine a social service system where the needy worked with a single caseworker who coordinated all services until they were returned to a productive role within society. Or more simply, imagine what would it be like to buy a car from a sales rep who was empowered to close the deal?

Empowerment and involvement lead to ownership and commitment. Within our organizations, co-workers would establish strong bonds and fierce alliances. Perhaps lifelong employment with an organization would again become the norm, keeping a core team together as the organization adapts. In the community, this commitment would lead to more intimate involvement in governance, something closer to a true democracy.

Blurred Boundaries

As we better appreciate our interdependence, our organizational boundaries will blur.

Imagine a society where employees work in multiple organizations. Through employee leasing and rotation arrangements, employees could relieve bottlenecks in their supplier organizations or work at a customer's site to better understand its needs. In some sectors, it might even be possible for employees to be shared between competitors. This practice is already common in architectural firms where trade secrets are minimal and workloads volatile. In this setting, it might be difficult for someone to answer the question, "Where do you work?"

What would it be like if employees worked across sectors? Business people could teach in schools. Teachers and students could work in industry. As these strategic alliances grew, opportunities for collaboration would develop, making better use of resources. For instance, schools might let a business use their facilities during nonschool hours in exchange for mentoring or access to the business's library. Perhaps a business would fund the construction of a research laboratory in exchange for some control over the research projects.

In our society, wasted resources are everywhere. With a more systemic view and stronger alliances, we all can gain.

HOW DO YOU CREATE A TOTAL QUALITY COMMUNITY?

Creating a total quality community requires assembling people from all sectors of society and getting them to collaborate. The challenges are enormous: turf, jargon, values. Where should we begin? Use these following suggestions as a guide.

Expand Community Involvement

Organizations must identify their distinctive competence not just for what they contribute to their marketplace but also to their community. Just as each organ in the body serves a special function, each organization should identify its unique skills, knowledge, or experience that could add value to society.

Volunteerism and community involvement must be directed toward strategic interdependencies. Just as 3M encourages employees to spend 15 percent of their time on innovations, organizations will want to encourage employees to spend some percentage of paid time working on strategic community projects through employee leasing or volunteering. Executive involvement on boards of directors should be based on the need to align strategies between sectors. Collaborative roundtables, where interdependent organizations forge self-imposed regulations for the good of all, will become commonplace. [4]

Not only should the organization step outside its boundaries, the community should step in.

Establishing organizational strategy should be viewed as a joint responsibility, including all stakeholder groups: employees, shareholders, regulators, customers, suppliers, and so on. Involving stakeholders can increase mutual trust and understanding while reducing the effort required to protect against litigation.

Catalyze Strategic Allies to Work Collaboratively

Organizations have interdependencies on two axes. Vertically, they are interdependent with their suppliers and customers. Horizontally, they are interdependent with other sectors of society. Organizations that want to catalyze community action (what we call "catalytic organizations") should map these interdependencies and then identify strategic alliances and key issues. Then they can work toward collaborative action.

Before collaboration will occur, three conditions must be satisfied:

- *Stake.* Organizations must believe that the stakes are high—either that the consequences of not collaborating are severe or that the opportunities afforded by collaborating are amplified.
- *Interdependence.* Organizations must believe that they need to work together to achieve success—that independent action is insufficient, and that they will be affected whether or not they participate.
- *Contribution.* Organizations must believe that they can affect the situation—that they have a valuable contribution to make.

Through an outreach strategy, catalytic organizations must educate their strategic allies on these conditions. Selling the problem or opportunity usually is preferable to selling solutions, so the catalytic organization should introduce TQM principles and techniques after the three conditions are met to avoid being branded a TQM zealot.

Once the allies feel a need to collaborate, the catalytic organization should provide a structure and process to facilitate their learning and collaboration. This may take the form of committees, roundtables, future search conferences, and the like.

Educate Allies on TQM Principles

As Einstein is believed to have said, "You cannot solve a problem with the thinking on the level that created it." Once a collaborative structure is in place, the catalytic organization should begin educating its allies on TQM and high-performance principles. Explore how common TQM strategies (such as benchmarking, reducing cycle time, and statistical process control) could be applied in different sectors.

Since the sectors use different jargon and come with different perspectives, a significant amount of time must be devoted to coming to common understandings. Documenting learning and definitions facilitates this process. Allies must be encouraged to view this learning as significant progress, for their task orientation may make them impatient for action. Structuring the meetings with clear learning outcomes will help the members feel more of a sense of achievement.

Transfer Ownership to the Stakeholders

The key issues and necessary changes cannot remain the agenda of a single organization. These must be shared by the stakeholders. So the catalytic organization should work toward collaborative action. For this to succeed, turfism must be set aside. Several tactics facilitate this collaboration.

First, establish written ground rules and values that will protect those who will be negatively impacted by change. Parties should come to the table with the assumption that there will be an important role for all to play and that everyone's role will change. Since the outcome is unknown and anxiety ridden, it will be important to establish a fair process in which all can have faith, just as we have faith in our democratic process even when we disagree with the results of an election.

Another tactic is to agree on a higher purpose. Rancorous debates often arise when issues are framed as two poles: protecting the environment or jobs, funding children or seniors, pursuing safety or profitability. As the two sides dig in their heels, both sink deeper in the quagmire. Instead, a catalytic leader will lead a dialogue to discover the common higher needs that all sides value. For instance, a small town in Washington state was faced with the dilemma of whether to let EPA violations force the closure of its primary employer, a copper smelter. At first, the citizens polarized, coming to meetings wearing buttons saying "Health" or "Jobs." After significant dialogue, the community began to question the wisdom of tying its future to one shaky industry. The higher purpose was quickly reframed as increasing economic diversity. Then citizens printed "Health *and* Jobs" buttons. [5]

Once consensus around a higher purpose is formed, organizations can be allowed freedom to approach it with different tactics. Groups that understand they are working toward the same goal are often more tolerant of experimentation and diverse approaches. Measuring the results of these different tactics can lead to important insights, while increasing appreciation for each other's contribution.

If the relationship between groups is so adversarial that no progress is likely, using a third organization to bridge the adversarial ones keeps disparate groups separate while linking them to an aligned agenda. [6]

Finally, share credit liberally. After all the hard work and behind-the-scenes action, a catalytic leader must be willing to watch others get the credit. For it is at this point that true ownership has been transferred. What Lao-Tse said is even more true now:

> A good leader talks little; but when his work is done, his aim filled, all others will say, "We did it ourselves."

CONCLUSION

The natural extension of TQM is to take it outside the organizational boundary and into the community. In this chapter, we explained why this move is important to our competitiveness, and we speculated on what a total quality community might be like. Finally, we provided some guidance about how to begin.

It is clear that creating the Total Quality Community will create enormous disruption to the status quo. However, if we stay true to the core values of delighting customers, empowering those on the front line, and finding ways for all to contribute, we can create a better world.

SUGGESTED READING

Bane, Mary, and David Ellwood. "Is American Business Working for the Poor?" *Harvard Business Review,* September–October 1991, pp. 58–66.

Gray, Barbara, and Donna Wood, eds. "Collaborative Alliances: Moving from Practice to Theory." *Journal of Applied Behavioral Science,* March 1991. Newbury Park: Sage Periodicals Press.

Luke, Jeff S. "Managing Interconnectedness: The Need for Catalytic Leadership." *Futures Research Quarterly,* Winter 1986, pp. 73–83.

Malone, M., and W. Davidow. "Virtual Corporation." *Forbes ASAP Technology Supplement,* December 7, 1992, pp. 102–07.

Morrison, Roy. *We Build the Road as We Travel.* Philadelphia: New Society Publishers, 1991.

Nirenberg, John. *The Living Organization.* Homewood, IL: Business One Irwin, 1993.

Schindler-Rainman, Eva, and Ronald Lippitt. *Building the Collaborative Community.* Riverside: University of California Extension, 1980.

Wheatley, Margaret. *Leadership and the New Science: Learning about Organizations from an Orderly Universe.* San Francisco: Berret-Koehler Publishers, 1992.

Whyte, William Foote, and Kathleen King Whyte. *Making Mondragon.* Ithaca, NY: ILR Press, 1988.

Notes

Chapter One Commitment

1. "Deploying TQM and Empowerment at McDonnell Douglas Space Systems Co." *Commitment Plus,* October 1992, p. 2.
2. Garvin, David A. "How the Baldrige Award Really Works." *Harvard Business Review,* November–December 1991, p. 90.

Chapter Two Justification and Timing

1. Stack, Jack. *The Great Game of Business: The Only Sensible Way to Run a Company.* New York: Currency Books, 1992, p. 7.

Chapter Three Education

1. Ginnodo, William L., and Richard Wellins. "Research Shows That TQM Is Alive and Well." *Tapping the Network Journal,* Winter 1992/1993, pp. 2–5.
2. Deming, W. Edwards. *Out of Crisis.* Cambridge, MA: MIT Center for Advanced Engineering Study, 1982, p. 42.
3. Odiorne, George. *Training by Objectives.* New York: Addison-Wesley Publishing, 1972, p. 106.
4. Crosby, Phillip B. *Completeness.* New York: New American Library/ Dutton, 1992, p. 18.

Chapter Four Results

1. Crosby, Philip B. "Viewpoint." *The Quality Observer,* May 1992, p. 2.
2. Gilbert, Thomas F. *Human Competence.* New York: McGraw-Hill, 1979, p. 40.
3. Lee, Chris. "Do the Job, Hold the Service." *Training Magazine,* November 1989, p. 8.

Chapter Seven Appraisals

1. Mitchell, T., and W. Silver. "Individual and Group Goals When Workers Are Interdependent: Effects on Task Strategies and Performance." *Journal of Applied Psychology* 75, no. 2 (1990), pp. 185–93.
2. Portions of this chapter are adapted from Hitchcock, Darcy. "Performance Management for Teams: A Better Way." *Journal for Quality and Participation: Association for Quality and Participation,* September 1990, pp. 52–57.
3. Pascale, Richard. *Managing on the Edge: How the Smartest Companies Use Conflict to Stay Ahead.* New York: Simon & Schuster, 1990.
4. Iaccoca, Lee, and William Novak. *Iaccoca: An Autobiography.* New York: Bantam Books, 1984.
5. Lawler, Edward III. *Strategic Pay: Aligning Organizational Strategies and Pay Systems.* San Francisco: Jossey-Bass, 1990.
6. Larson, Carl E., and Frank M. J. La Fasto. *Team Work: What Must Go Right/What Can Go Wrong.* Newbury Park, CA: Sage Publications, 1989.
7. Hoshin is a Japanese planning process geared toward performance breakthroughs. For more information, see King, Bob. *Hoshin Planning: The Developmental Approach.* Methuen, MA: Goal/QPC, 1989.

Chapter Eight Rewards

1. Lawler, Edward E. III. *Strategic Pay.* San Francisco: Jossey-Bass, 1990.
2. Schroeder, Les. *Training Magazine,* September 1990, p. 70.
3. O'Brien, Richard M., et al. *Industrial Behavior Modification.* Elmsford, NY: Pergamon Press, 1978, p. 214.
4. Robertson, R. N., and C. I. Osuorah. "Gainsharing in Action at Control Data." *Journal of Quality and Participation,* December 1991, p. 28.
5. Peters, Tom. *Thriving on Chaos.* New York: Alfred A. Knopf, 1988, p. 339.

Chapter Nine Power Structure

1. Lawler, Edward E. III. *High Involvement Management.* San Francisco: Jossey-Bass, 1986.
2. Macy; Izumi; Bliese; and Norton. "Organizational Change and Work Innovation: A Meta-Analysis of 131 North American Field Studies 1961–1991." *Research in Organizational Change and Development,* vol. 7, ed. Woodman and Passmore. Greenwich, CT: JAI Press, 1993.

3. Iaccoca, Lee, and William Novak. *Iaccoca: An Autobiography.* New York: Bantam Books, 1984.
4. Lawler. *High Involvement Management.*
5. For information about the demonstrated benefits of these levels of empowerment, see Lawler. *High Involvement Management.*
6. Commission on the Skills of the American Workforce. *America's Choice: High Skills or Low Wages!* Washington, National Center of Education and the Economy, June 1990. Statistics: Managers = 22%; professionals = 27%; technicians and supervisors = 8%; professional sales = 9%; noncollege-educated technicians and supervisors = 11%; clerical = 4%; laborers = 5%; skilled crafts = 9%; and retail clerks = 5%.
7. Lawler, Edward, Monty Mohrman, and Jerry Ledford. *Employee Involvement and Total Quality Management: Practices and Results in Fortune 1000 Companies.* San Francisco: Jossey-Bass, 1992.
8. For an introduction to the topic of sociotechnical systems design, see Hitchcock, Darcy, and Lord, Linda, "The New Convert's Primer to Socio-Tech." *Journal for Quality and Participation,* June 1992, pp. 46–57.
9. The results of this study were published in Hitchcock, Darcy. "Overcoming the Top Ten Self-Directed Team Stoppers." *Journal for Quality and Participation* (Association for Quality and Participation), December 1992, pp. 42–47. Portions of this section are adapted from that article.
10. For information on creative compensation options, see Bureau of National Affairs. "Changing Pay Practices: New Developments in Employee Compensation." Washington, DC: *BNA Labor Relations Week* 2, no. 24 (June 15, 1988).
11. Linda Moran, executive for Zenger-Miller. See Geber, Beverly. "From Manager into Coach." *Training Magazine,* February 1992, pp. 25–31.
12. Miller, Lawrence. *American Spirit: Visions of a New Corporate Culture.* New York: William Morrow, 1984.
13. Carnevale, Anthony. *America and the New Economy.* Washington, DC: American Society for Training and Development and U.S. Department of Labor, 1991.
14. Excerpted from Hitchcock and Lord. "The New Convert's Primer to Socio-Tech."

Chapter Ten Management Beliefs

1. Pascale, Richard Tanner. *Managing on the Edge: How the Smartest Companies Use Conflict to Stay Ahead.* New York: Simon & Schuster, 1990.

2. Fisher, K. Kim. "Managing in the High-Commitment Workplace." *Organizational Dynamics,* Winter 1989.
3. Ibid.
4. Crosby, Philip B. *Quality without Tears: The Art of Hassle-Free Management.* New York: McGraw-Hill, 1984.
5. Gabel, Natalie. "Is 99.9% Good Enough?" *Training Magazine,* March 1991, pp. 40–41.
6. Johnson, H. Thomas. *Relevance Regained: From Top-Down Control to Bottom-Up Empowerment.* New York: Free Press, 1992. Reprinted with permission of The Free Press, a Division of Macmillan, Inc. Copyright © by H. Thomas Johnson.
7. Stayer, Ralph. "How I Learned to Let My Workers Lead." *Harvard Business Review,* November–December 1990, p. 66(10).
8. Albrecht, Karl, and Ron Zemke. *Service America! Doing Business in the New Economy.* Homewood, IL: Dow Jones Irwin, 1985.
9. Miller, Lawrence M. *American Spirit: Visions of a New Corporate Culture.* New York: William Morrow, 1984.
10. The Four Cs are a trademark of AXIS Performance Advisors, Inc.
11. Geber, Beverly. "From Manager to Coach." *Training Magazine.* February, 1992, pp. 25–31.

Chapter Eleven Structure

1. Carnevale, Anthony. *America and the New Economy.* Washington, DC: American Society for Training and Development and U.S. Department of Labor, 1991.
2. Pinchot, Gifford III. *Intrapreneuring: Why You Don't Have to Leave a Corporation to Become an Entrepreneur.* New York: Harper & Row, 1986.
3. Kanter, Rosabeth Moss. *When Giants Learn to Dance: Mastering the Challenge of Strategy, Management, and Careers in the 1990s.* New York: Simon & Schuster, 1989.
4. Ibid.
5. The Eight Ss are a trademark of AXIS Performance Advisors, Inc.
6. Pascale, Richard Tanner. *Managing on the Edge: How the Smartest Companies Use Conflict to Stay Ahead.* New York: Simon & Schuster, 1990. Copyright © 1990 by Richard Pascale. Reprinted by permission of Simon & Schuster, Inc.
7. Prahalad, C. K., and Gary Hamel. "The Core Competence of the Corporation." *Harvard Business Review,* May–June 1990, pp. 79–91.

8. Pascale, Richard Tanner. *Managing on the Edge. How the Smartest Companies Use Conflict to Stay Ahead.* New York: Simon & Schuster, 1990. Copyright © by Richard Pascale. Reprinted by permission of Simon & Schuster, Inc.
9. Ibid.
10. Osborne, David, and Ted Gaebler. *Reinventing Government: How the Entrepreneurial Spirit Is Transforming the Public Sector.* New York: Addison-Wesley Publishing, 1992.
11. Ibid.
12. Interestingly, after 1.5 years, they recreated a receptionist position, but now everyone takes responsibility to make the position humane.
13. Johnson, H. Thomas. *Relevance Regained: From Top-Down Control to Bottom-Up Empowerment.* New York: Free Press, 1992.

Chapter Twelve Systems

1. Osborne, David, and Ted Gaebler. *Reinventing Government: How the Entrepreneurial Spirit Is Transforming the Public Sector.* New York: Addison-Wesley Publishers, 1992.
2. Johnson, H. Thomas. *Relevance Regained: From Top-Down Control to Bottom-Up Empowerment.* New York: Free Press, 1992. Reprinted with permission of The Free Press, a Division of Macmillan, Inc. Copyright © by H. Thomas Johnson.
3. Ibid.
4. Freedman, David. "An Unusual Way to Run a Ski Business." *Forbes ASAP Technology Supplement,* December 7, 1992, pp. 27–32.
5. LaPlante, Alice. "Shared Destinies: CEOs and CIOs." *Forbes ASAP Technology Supplement,* December 7, 1992, pp. 32–42.
6. This case is adapted from Hitchcock, Darcy. "The Engine of Empowerment." *Journal for Quality and Participation,* March 1992, pp. 50–58.
7. Lawler, Edward III. *Strategic Pay: Aligning Organizational Strategies and Pay Systems.* San Francisco: Jossey-Bass, 1990.
8. LeBoeuf, Michael. "The Greatest Management Principle in the World." *Working Woman,* January 1988, pp. 70–74.

Chapter Thirteen Organizational Learning

1. Senge, Peter. "The Leader's New Work: Building Learning Organizations." In John Renesch, ed. *The New Traditions in Business.* San Francisco: Sterling & Stone, 1991.

2. Carnevale, Anthony, and Leila Gainer. *The Learning Enterprise.* Washington, DC: The American Society for Training and Development and the U.S. Department of Labor, 1989, p. 3.
3. Interview with John Coné, director of Sequent University. Sequent Computer Systems. February 16, 1993.
4. Leonard-Barton, Dorothy. "The Factory as a Learning Laboratory." *Sloan Management Review,* Fall 1992, pp. 23–38.
5. Ibid.
6. Pascale, Righard T. *Managing on the Edge: How the Smartest Companies Use Conflict to Stay Ahead.* New York: Simon & Schuster, 1990.
7. Osborne, David, and Ted Gaebler. *Reinventing Government.* Reading, MA: Addison-Wesley Publishing, 1992.
8. Weisbord, Marvin R. *Productive Workplaces.* San Francisco: Jossey-Bass, 1987, p. 95.
9. Senge, Peter. *The Fifth Discipline.* New York: Doubleday, 1990, p. 287.
10. PartNURShip is a registered trademark of Southwest Washington Medical Center.
11. Senge. *The Fifth Discipline.*
12. Drucker, Peter. "Japan: New Strategies for a New Reality." *The Wall Street Journal,* October 2, 1991.
13. Drucker. *Management.* New York: Harper & Row, 1974, p. 428.
14. Senge, *The Fifth Discipline.*
15. Wick, Calhoun, and Lu Stanton Leon. *The Learning Edge.* New York: McGraw-Hill, 1993.
16. Carnevale, Anthony, and Leila Gainer. *Workplace Basics: The Skills Employers Want.* Washington, DC: The American Society for Training and Development and the U.S. Department of Labor, 1989.
17. Senge. *The Fifth Discipline.*
18. Ibid.
19. Argyris, Chris. "Teaching Smart People How to Learn." *Harvard Business Review,* May/June 1991, pp. 99–109.
20. Ibid.
21. Peters, Tom. "Curiosity Killed the Company? It's More Likely Dullness Was the Cause." *Chicago Tribune,* November 11, 1991.
22. Senge. *The Fifth Discipline.*
23. Hitchcock, Darcy; Marsha Willard; and Ed Warnock. *The Catalytic Leadership Handbook.* Portland, OR: Portland State University, Partners for Human Investment, 1993.

Epilogue The Future

1. Osborne, David, and Ted Gaebler. *Reinventing Government.* Reading, MA: Addison-Wesley Publishing, 1992.
2. *The Oregonian,* Section H, April 4, 1993, p. 1.
3. See books by Morrison and Whyte in Suggested Reading for more information.
4. For an example of this, see the environmental case study of high-tech firms in the San Jose area (where they collaborated to devise the Model Hazardous Materials Storage Permit Ordinance, which became the basis for state and federal standards) published in: Logsdon, Jeanne. "Interests and Interdependence in the Formation of Social Problem-Solving Collaborations." *Journal of Applied Behavioral Science, Collaborative Alliances*, Part 1 (March 1991), pp. 23–37.
5. Heifetz, Ronald. *Leadership With and Without Authority* (title tentative). Cambridge, MA: Harvard University Press, forthcoming 1994.
6. Westley, Frances, and Harrie Vredenburg. "Strategic Bridging: The Collaboration between Environmentalists and Business in the Marketing of Green Products." *Journal of Applied Behavioral Sciences, Collaborative Alliances,* Part 1 (March 1991), pp. 65–90.

Index

T

THE SERVICE/QUALITY SOLUTION

Using Service Management to Gain Competitive Advantage

David A. Collier

Co-published with ASQC Quality Press

Improve your service strategy and survive the pressures within today's marketplace with Collier's 16 tools for effective service/quality management.

ISBN: 1-55623-753-7

GLOBAL QUALITY

A Synthesis of the World's Best Management Methods

Richard Tabor Greene

Co-published with ASQC Quality Press

This comprehensive resource organizes the chaos of quality improvement techniques so you can identify the best approaches for your organization. Includes the 24 quality approaches used worldwide, the essentials of process reengineering, software techniques, and seven new quality improvement techniques being tested in Japan.

ISBN: 1-55623-915-7

Available at bookstores and libraries everywhere.

Please Send More Information About

- ❑ PARS
- ❑ Membership
- ❑ Education Courses
- ❑ Conferences
- ❑ *The Journal for Quality and Participation*

Name____________________ Title____________________

Company____

Address____

City, State, Zip__________________________________

Phone____________________ Fax____________________

BUSINESS REPLY MAIL

FIRST CLASS MAIL PERMIT NO. 16494 CINCINNATI OH

Postage Will Be Paid By Addressee

AQP
ASSOCIATION FOR QUALITY AND PARTICIPATION
801 B WEST 8TH STREET
CINCINNATI OH 45203-9946